Simon Haykin

DFT/FFT and Convolution Algorithms

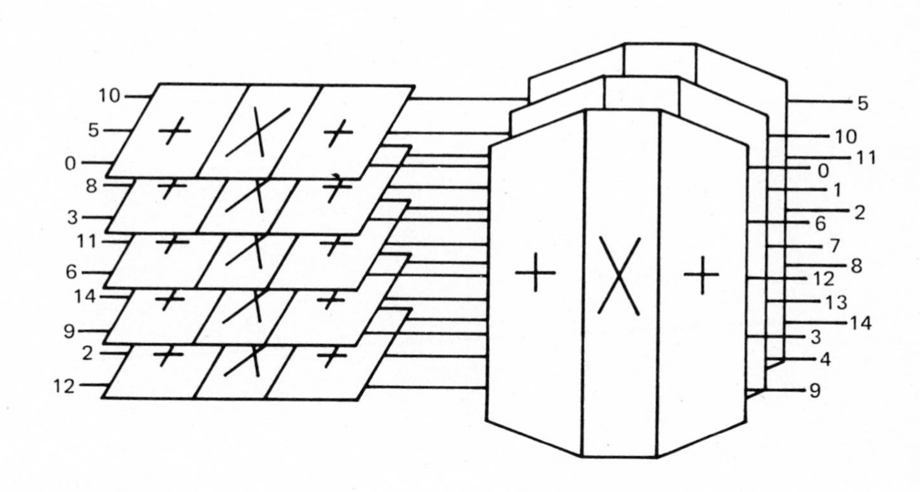

TOPICS IN DIGITAL SIGNAL PROCESSING

C. S. BURRUS and T. W. PARKS: *DFT/FFT AND CONVOLUTION ALGORITHMS*
Rice University

T. W. PARKS and C. S. BURRUS: *DIGITAL FILTER DESIGN* (in preparation)
Rice University

W. KOHN: *DIGITAL CONTROL* (in preparation)

J. TREICHLER, M. LARIMORE, and R. JOHNSON, JR.: *A PRACTICAL GUIDE TO ADAPTIVE FILTER DESIGN*
Applied Signal Technology, Inc., Argosystems, and Cornell University

DFT/FFT and Convolution Algorithms

THEORY AND IMPLEMENTATION

C. S. BURRUS and T. W. PARKS

Department of Electrical and Computer Engineering

Rice University

with TMS32010 programs by

James F. Potts

Texas Instruments

A Wiley-Interscience Publication

JOHN WILEY & SONS

New York • Chichester • Brisbane • Toronto • Singapore

The authors acknowledge the support of Texas Instruments, Inc., and the National Science Foundation for part of the work presented here.

Library of Congress Cataloging in Publication Data:

Burrus, C.S.
DFT/FFT and convolution algorithms.

(Topics in digital signal processing)
"A Wiley-Interscience publication."
Includes index.
1. Signal processing—Digital techniques. 2. Fourier transformatons. 3. Convolutions (Mathematics) I. Parks, T.W. II. Potts, James F. III. Title, IV. Title: D.F.T./F.F.T. and convolution algorithms. V. Series.
TK5102.5.B77 1984 621.38'0433 84-18808
ISBN 0-471-81932-8

10 9 8 7 6 5 4 3

To

Mary and Martha

PREFACE

This algorithm book has been written for the scientist or engineer who has a good understanding of continuous-time signals and who has been introduced to discrete-time signal analysis.

The main topic of this book is the Fourier transform of a discrete-time signal, with emphasis on efficient algorithms for computing the Discrete Fourier Transform (DFT). Although one might want to call this an "FFT Handbook," the main theme of this book is that there is no such thing as *the* FFT. Rather, there is a collection of ideas and approaches to computing the DFT. Since technology is changing so rapidly in this area, it is important to understand the fundamental approaches to efficient computation. While the radix-4 , length-64, Cooley-Tukey FFT may be the best approach for the Texas Instruments TMS32010, there may be better choices for future signal processing chips as technology evolves.

The first chapter reviews continuous- and discrete-time transform analysis of signals and discusses sampling theory. Chapter 2 begins with a presentation of properties of the DFT. After a discussion of several ways to compute the DFT at a few frequencies (direct, Goertzel, and chirp transforms), the three main approaches to an FFT (Cooley-Tukey, prime-factor, and Winograd transforms) are described in considerable detail. FORTRAN statements are used to explain concepts. Chapter 3 covers both linear and circular convolution of discrete-time signals, and includes a discussion of efficient ways to compute a convolution. Chapter 4 contains a collection of FORTRAN programs for the DFT which may be used directly or may be used as a basis for custom program development for special applications. These programs use floating-point arithmetic and are intended for general-purpose computers. Chapter 5 has a collection of programs for the TMS32010 signal processing chip, using fixed-point arithmetic. Chapter 6 concludes the book with a comparison of the different algorithms that have been programmed. Good algorithms for the TMS32010 chip are recommended.

There are several ways this book may be used. For some applications, one might simply turn to Chapter 5 and copy the macro-assembler statements for a radix-2 Cooley-Tukey FFT (one of the shorter programs in Chapter 5). If this program is too slow or if it uses too much memory, one should read the recommendations in Chapter 6. If the solution is still not found, one should dig deeper into the theory presented in this book or in the suggested references to develop his own customized program.

The individual chapters have been written to be self-contained. The first three chapters, in particular, contain some repetition of material to make it easy for the reader to jump to the middle of the book, rather than reading the material in sequence.

This book could not have been written without the support of Texas Instruments, Inc. We would especially like to acknowledge the support and encouragement of John Hayn, who suggested that we write this book and made it easy for us to do so. Maridene Lemmon has refined countless versions of the text and improved our writing with her patient advice. Mike Hames and Surendar Magar have helped us understand the TMS32010. Jim Potts, with help from Ray Simar, has written the programs for the TMS32010. Jim has been ready to help whenever we encountered a problem. He has provided the data for timing and evaluation of the various algorithms which have been run on the TMS32010.

We would also like to thank Professor H.W. Schuessler and our graduate students, Doug Jones and David Scheibner, who carefully read the text and made several suggestions for improvements.

C. S. BURRUS
T. W. PARKS

Houston, Texas
June 1984

CONTENTS

ILLUSTRATIONS

FIGURE PAGE

TABLES

TABLE PAGE

DFT/FFT and Convolution Algorithms

Chapter 1

FREQUENCY-DOMAIN DESCRIPTION OF SIGNALS

A review of some basic properties of the Fourier series and Fourier transform methods of representing signals is presented in this chapter. The purpose is to establish a point of view and introduce some notation to be used throughout the book. This notation is consistent with the standards set up by the IEEE [1].

Key features of Fourier analysis of continuous-time signals, which carry over into the analysis of discrete-time signals, are reviewed in this chapter. Fourier transforms of limited data are discussed, along with windowing and signal modelling. The theory of sampling continuous-time signals to produce discrete-time signals is presented. Fourier analysis of discrete-time signals is developed and related to Fourier analysis of continuous-time signals. The first chapter concludes with a comparison of continuous and discrete-time transforms and a brief discussion of convolution.

The continuous-time signals described in this book are usually represented by functions of a real variable, "t", considered to represent time. The variable "t" could also represent distance "z", or some other variable. In applications such as image processing or sonar and seismic signal analysis, the signals depend on several variables (i.e., time and two or three spatial variables in seismic processing). While this book focuses on functions of one time (or space) variable, many of the concepts and algorithms developed here can be extended to signals which are functions of several variables (multidimensional signals).

The discrete-time signals in this book are real or complex-valued functions defined on the integers. A discrete-time signal "x" is written either as x_n or as x(n). The index "n" could, of course, represent distance or even the day of the week. While this book concentrates on discrete-time signals which are functions of one integer, most of the ideas and algorithms can be used for discrete-time signals which are functions of several integers (multidimensional discrete-time signals). For example, a two-dimensional DFT can be computed by using the one-dimensional FFT algorithms (described in this book) on rows and columns of the two-dimensional array.

1.1 CONTINUOUS-TIME SIGNALS

In continuous-time signal analysis, signals are normally represented as real valued functions of the real variable t. The expression of a signal as s(t) represents the entire function for all values of t in the domain of definition of the function, not just the real number which is the value of the function for a particular time t.

It is often more efficient or more revealing to deal with some representation of the signal other than the function s(t) itself. Generally, in the processing or analysis of a signal, an attempt is made to better extract or describe some property or characteristic not easily seen from the function s(t) itself. In other words, a different coordinate set or description is sought that will better show the desired property. Obviously, no information can be created, but it can be better displayed.

The Fourier signal representations described in this chapter make use of weighted combinations of elementary signal components with different frequencies to expose the frequency content of signals. If a signal is passed through a linear time-invariant system, the frequency components are changed only in magnitude and phase, but not in frequency. The effect of the system on the frequency spectrum is much easier to predict and interpret than the effect on the time function s(t) (see Section 1.4). In other cases, the Fourier coefficients contain information in more useable form or in a more compact form. For example, quantized Fourier coefficients are used in data compression to give more efficient transmission and storage of speech signals.

The description of frequency-domain representation of signals begins with the Fourier series in which the signal is represented as a summation of a countable number of frequency components with equal frequency increments (a discrete-frequency representation). The Fourier transform is then described as a continuous-frequency representation of the signal using an integral superposition of elementary signals.

An example using a square time pulse is used throughout the discussion of the Fourier series and transform to clearly illustrate the similarites and differences of the two Fourier frequency-domain signal representations (see Examples 1-1a, 1-1b, and 1-2).

1.1.1 The Fourier Series

The Fourier series may be viewed as a way to represent a signal s(t) as a superposition of harmonically related sine and cosine waves [2]. For example, the square pulse p(t), shown as an even function of time in Figure 1-1, may be represented as a superposition of cosines on the time interval of length T.

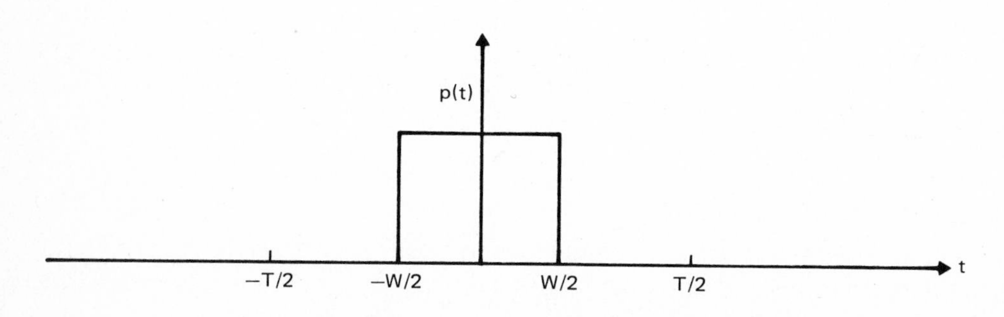

FIGURE 1-1. THE SQUARE PULSE

For this even function, p(t), the Fourier series has only cosine terms. The first three terms in the series are

$$\bar{p}(t) = a_0 + a_1 \cos(2\pi t/T) + a_2 \cos(4\pi t/T) \quad (1.1)$$

Since the cosines are orthogonal on the interval (−T/2,T/2), the coefficients are calculated as

$$a_k = 1/T \int_{-T/2}^{T/2} p(t) \cos(2\pi kt/T)\, dt \quad (1.2)$$

The square pulse may not be centered at the origin but may be shifted to the right, as shown in Figure 1-2.

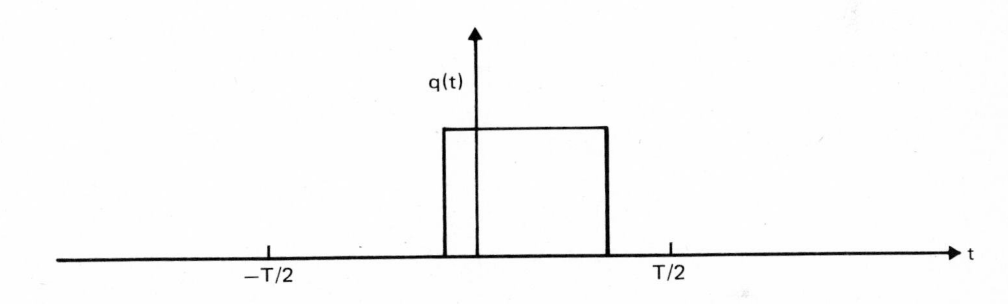

FIGURE 1-2. A SHIFTED SQUARE PULSE

Both sine and cosine terms are needed in this three- frequency expansion. The expansion is

$$\bar{q}(t) = a_0 + a_1 \cos(2\pi t/T) + a_2 \cos(4\pi t/T) + b_1 \sin(2\pi t/T) + b_2 \sin(4\pi t/T) \quad (1.3)$$

Instead of this sine and cosine notation, the more compact complex exponential representation is often used, especially when there are many harmonics in the Fourier expansion. A three-frequency expansion is

$$\bar{q}(t) = c_0 + c_1 e^{j2\pi t/T} + c_2 e^{j4\pi t/T} + c_{-1} e^{-j2\pi t/T} + c_{-2} e^{-j4\pi t/T} \quad (1.4)$$

The Fourier series coefficients for a signal x(t), using this complex number notation, are given by

$$c_k = 1/T \int_{-T/2}^{T/2} x(t)\, e^{-j2\pi kt/T}\, dt \tag{1.5}$$

As more and more frequencies are included in the approximation, the error tends to zero for well-behaved (i.e., finite energy) signals, and

$$x(t) = \sum_{k=-\infty}^{\infty} c_k\, e^{j2\pi kt/T} \tag{1.6}$$

Since the expansion (1.6) is a periodic function of time t with a period T, (1.6) may be considered to be the Fourier expansion of the signal $f_p(t)$ with period T expressed as

$$f_p(t) = \sum_{m=-\infty}^{\infty} f_r\,(t - mT) \tag{1.7}$$

where

$$f_r(t) = \begin{cases} x(t) \text{ for } t \in [-T/2, T/2] \\ 0 \text{ elsewhere} \end{cases} \tag{1.8}$$

Thus, the Fourier series is understood in two ways. It is either viewed as an expansion of f(t) on a finite time interval $(-T/2, T/2)$, or as an expansion of a periodic signal with period T.

Summary of Fourier Series Characteristics

1) The set of complex coefficients c_k in (1.5) is called the spectrum of the signal x(t).
2) When the signal x(t) is real, the c_k have a conjugate symmetry:

$$c_{-k} = c_k^* \tag{1.9}$$

The spectrum of a real signal has even magnitude and odd phase

$$|c_{-k}| = |c_k| \text{ and } \measuredangle\, c_{-k} = -\measuredangle\, c_k \tag{1.10}$$

3) The truncated Fourier series, expressed as

$$\bar{x}(t) = \sum_{k=-N}^{N} c_k\, e^{j2\pi kt/T} \tag{1.11}$$

is the best approximation to x(t) in the sense that the integrated squared error

$$E = \int_{-T/2}^{T/2} |x(t) - \bar{x}(t)|^2 \, dt \tag{1.12}$$

is minimized.

4) Parseval's theorem states that

$$1/T \int_{-T/2}^{T/2} x^2(t) \, dt = \sum_{k=-\infty}^{\infty} |c_k|^2 \tag{1.13}$$

This theorem means that the energy of a signal may be calculated in the time domain or in the frequency domain. Further, the amount of energy in a given band of frequencies may be found by summing over only those index values corresponding to the frequency band of interest.

$$K_1 \leq k \leq K_2 \tag{1.14}$$

This review of the Fourier series concludes with an example consisting of two parts: a Fourier series for a square pulse and a Fourier series for a square wave.

Example 1-1a: Fourier Series for a Square Pulse

According to (1.5) and (1.6), the coefficients for the Fourier expansion on the interval $-T/2, T/2$ of the square pulse, shown in Figure 1-1, are given by

$$c_k = 1/T \int_{-T/2}^{T/2} p(t) \, e^{-j2\pi kt} \, dt = 1/T \int_{-W/2}^{W/2} e^{-j2\pi kt/T} \, dt \tag{1.15}$$

or

$$c_k = W/T \, \frac{\sin(\pi kW/T)}{(\pi kW/T)} \tag{1.16}$$

The Fourier coefficients in this example are real, because the signal is an even function of time (see Figure 1-3.)

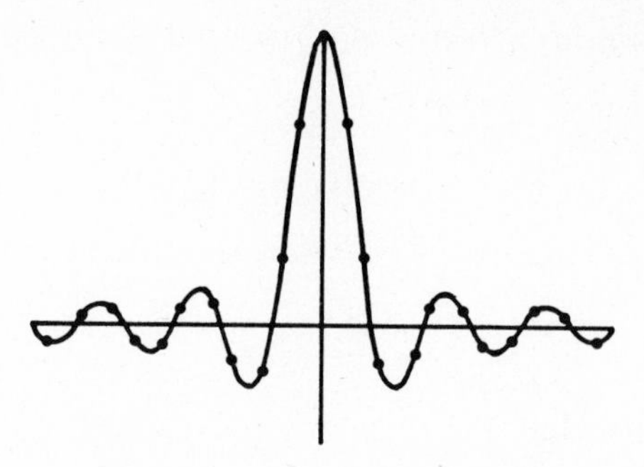

FIGURE 1-3. FOURIER SPECTRUM FOR A SQUARE PULSE

Example 1-1b: Fourier Series for a Square Wave

The Fourier expansion of the square wave, shown in Figure 1-4, may be obtained from the Fourier expansion of the square pulse in Example 1-1a by using (1.7) and (1.8).

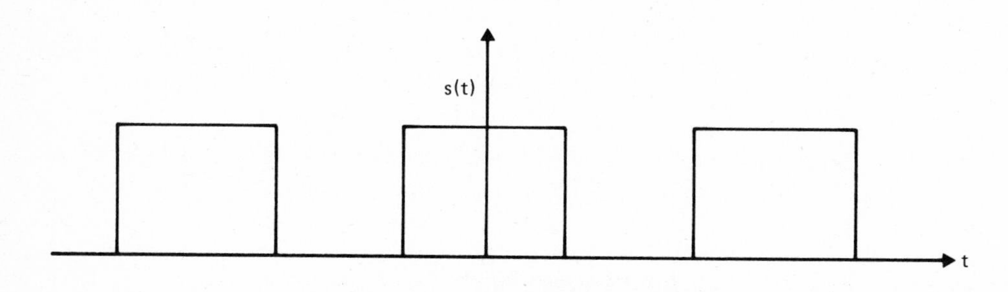

FIGURE 1-4. A SQUARE WAVE

The Fourier series for this periodic signal with period T is exactly the same as for the square pulse in Example 1-1a, since s(t) is simply a periodic extension, with period T, of the square pulse.

1.1.2 The Fourier Transform

As in the case of the Fourier series, the Fourier transform in the form of

$$S(\omega) = \int_{-\infty}^{\infty} s(t)\, e^{-j\omega t}\, dt \tag{1.17}$$

and the inverse Fourier transform in the form of

$$s(t) = 1/2\pi \int_{-\infty}^{\infty} S(\omega)\, e^{j\omega t}\, d\omega \tag{1.18}$$

may be viewed as an expansion of a signal as a superposition of exponentials [3]. In this case, however, the frequencies take on all real values rather than simply the harmonics (discrete frequencies). The Fourier transform is a "continuous-frequency" signal representation with no requirement that the signal be periodic. In fact, the Fourier integral in (1.17) converges for finite energy signals, which cannot be periodic. The Fourier series is used for finite power periodic signals, while the Fourier transform is used for finite energy signals, defined for all time.

If a signal s(t) is time-limited, i.e., if s(t) = 0 for t outside the time interval (−T/2, T/2), then the Fourier series coefficients of this signal may be obtained by sampling the Fourier transform, shown as

$$c_k = \frac{1}{T} S(\omega) \Big|_{\omega = k2\pi/T} \qquad (1.19)$$

Summary of Fourier Transform Characteristics

1) The complex valued function $S(\omega)$ of the real frequency variable ω, as defined in (1.17), is called the (Fourier) spectrum of the signal s(t).

2) When the signal s(t) is real, the spectrum has a conjugate symmetry:

$$S(-\omega) = S(\omega)^* \qquad (1.20)$$

The spectrum of a real signal s(t) has even magnitude $|S(\omega)|$ and odd phase $\sphericalangle S(\omega)$.

3) The truncated (bandlimited) inverse Fourier transform written as

$$\bar{s}(t) = 1/2\pi \int_{-B}^{B} S(\omega)\, e^{j\omega t}\, d\omega \qquad (1.21)$$

is the best approximation to s(t) using frequencies limited to the band (−B,B) in the sense that the integrated squared error, shown as

$$E = \int_{-\infty}^{\infty} |s(t) - \bar{s}(t)|^2\, dt \qquad (1.22)$$

is minimized.

4) Parseval's theorem states that

$$\int_{-\infty}^{\infty} |s(t)|^2 \, dt = 1/2\pi \int_{-\infty}^{\infty} |S(\omega)|^2 \, d\omega \tag{1.23}$$

with the same implications as for the Fourier series.

5) The value of the Fourier transform at any one particular frequency $S(\omega_o)$ is a linear functional of the signal s(t). This means that the entire signal s(t) must be available before the spectrum can be calculated. This characteristic of the Fourier transform has important implications for real-time spectral analysis and is discussed in more detail in Section 1.1.3.

This brief review of the Fourier transform concludes with an example of a Fourier transform of a square pulse. The square pulse can be analyzed in terms of either a Fourier series expansion over a finite time interval (Example 1-1a), or a Fourier transform over infinite time. The Fourier transform can be regarded as an extension of the Fourier series as the finite time interval tends to ∞.

Example 1-2: Fourier Transform of a Square Pulse

This example demonstrates calculation of the Fourier transform of the square pulse shown in Figure 1-1. This Fourier transform is compared with the Fourier series for a square pulse, calculated in Example 1-1.

From (1.17), the Fourier transform is

$$\begin{aligned} S(\omega) &= \int_{-\infty}^{\infty} p(t)\, e^{-j\omega t} \, dt = \int_{-W/2}^{W/2} e^{-j\omega t} \, dt \\ &= \frac{W\sin(\omega W/2)}{(\omega W/2)} \end{aligned} \tag{1.24}$$

The Fourier transform of this signal is real, because the signal is an even function of time (see Figure 1-5).

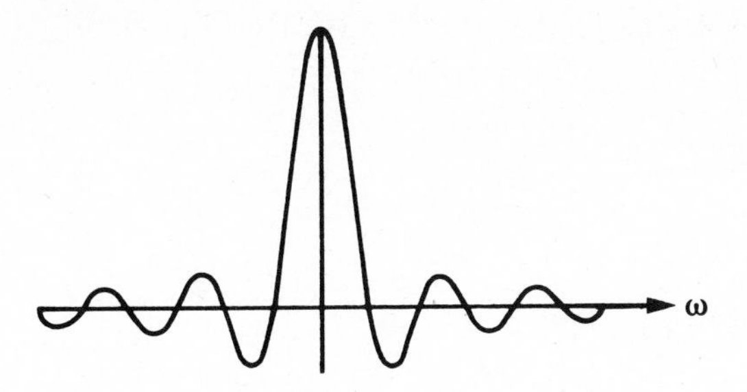

FIGURE 1-5. FOURIER TRANSFORM OF A SQUARE PULSE

Since this signal is time-limited, the Fourier series coefficients can be found by sampling the Fourier transform. Thus:

$$S(\omega)\Big|_{\omega = 2\pi k/T} = \frac{W \sin(\pi k W/T)}{(\pi k W/T)} \tag{1.25}$$

is the same function, except for the constant factor 1/T, as obtained in Example 1-1 using the Fourier series. The Fourier series is a signal representation using frequencies spaced 1/T Hz or $2\pi/T$ radians/second apart.

1.1.3 Windows, Segmentation, and Signal Models

The entire signal s(t) must be known for *all* values of time in order to calculate the Fourier transform using (1.17). In order to calculate the Fourier series using (1.5), the values of the signal must be known for all time t in the interval $-T/2, T/2$. If only a short interval of a time function is known, the Fourier spectrum of the signal cannot be found without making some assumptions about the behavior of the signal outside the short interval.

There are several standard assumptions about the signal behavior. The segment which has been measured can be periodically extended. The signal can be assumed to be identically zero outside the measurement interval. An extrapolation of the signal can be made outside of the interval based on some model of the signal. One typical model assumes the signal is bandlimited to a known band of frequencies. Another family of modelling techniques assumes the signal is a finite sum of exponential time functions and proceeds to find the unknown exponentials in this parametric model. Most of these signal extrapolation techniques are described in the spectral estimation literature [4].

In the spectral analysis of non-stationary signals such as speech, the signal is multiplied by a window function to isolate a certain time epoch. The multiplication by this window corresponds to a convolution of the underlying signal's Fourier transform with the Fourier transform of the window resulting in a smearing or broadening of the spectrum.

While the main emphasis of this book is on the algorithms used to compute Fourier transforms, the reader should be aware that the basic problem of estimating a Fourier spectrum from a segment of data involves assumptions, often implicit, about a signal model. A thorough understanding of these concepts is necessary for successful application of the algorithms in this book.

1.2 DISCRETE-TIME SIGNALS

1.2.1 Sampling Continuous-Time Signals

A discrete-time signal is often obtained from a continuous-time signal by sampling at equally spaced intervals in time. If the discrete-time signal is denoted by s_n, then the sampling of s(t) every T seconds gives

$$s_n = s(nT) \qquad n = \ldots -1, 0, 1, 2, \ldots \tag{1.26}$$

There is a fundamental ambiguity as to which continuous-time signal is being represented by a given set of samples. This may be explained with the following example:

Example 1-3: Ambiguity (Aliasing) when Sampling a Cosine Wave

If the samples of a cosine wave are obtained at a rate of five samples per second, shown as

$$s_n = 1, -.81, .31, .31, -.81, 1, -.81, .31, \ldots \tag{1.27}$$

it is reasonable to assume a sampling of a 2-Hz signal, shown as a solid line in Figure 1-6. It is also possible to obtain exactly the same set of samples by sampling the 3-Hz cosine, shown as a dashed line in Figure 1-6. In fact, there are infinitely many cosine waves at different frequencies which would give exactly the same samples!

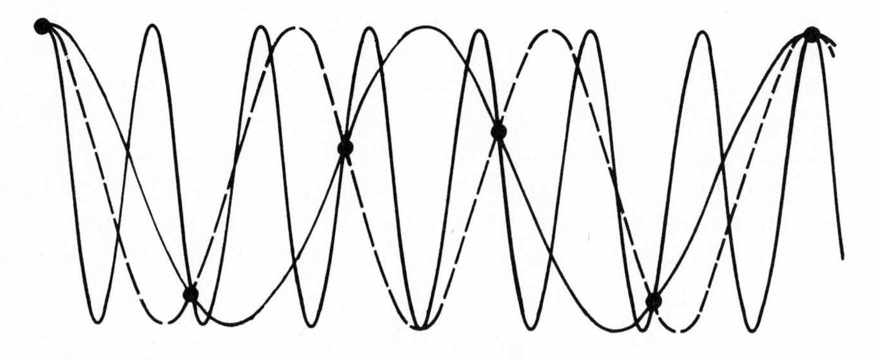

FIGURE 1-6. ALIASING

The usual convention used to relate a unique signal to a set of samples is to assume that the signal is a lowpass bandlimited signal. As long as the sampling rate is higher than twice the highest frequency present in the signal, there is a unique continuous-time signal associated with a discrete-time signal, viewed as a set of samples. In Example 1-3, the sampling rate is five samples per second. As long as the highest frequency in the signal is less than 2.5 Hz, there is a unique continuous-time signal associated with the sample set. In Example 1-3, a 2-Hz cosine is related to the given sample set. The phenomenon of having other names for continuous-time signals with the same samples is called aliasing. At this point, the following sampling theorem can be stated:

SAMPLING THEOREM:

> A bandlimited signal, with highest frequency B Hz, can be uniquely recovered from its samples as long as the sampling rate is higher than 2B samples per second. Lower sampling rates may result in aliasing.

The sampling theorem stated above assumes that infinitely many time samples are available for reconstruction of the sampled signal. The reconstruction or interpolation of the signal is an infinite sum of interpolation functions weighted by the time samples. When $T = 1/2B$,

$$x(t) = \sum_{-\infty}^{\infty} x(nT)\,\mathrm{sinc}(t-nT) \tag{1.28}$$

where

$$\mathrm{sinc}(t) = \frac{\sin(\pi t/T)}{\pi t/T} \tag{1.29}$$

In practice, it is not possible to obtain infinitely many samples so an approximate interpolation must be made. A higher sampling rate than that dictated by the "sampling theorem" should be used. The approximate interpolation, using N measured samples, has a form

$$\bar{x}(t) = \sum_{n=0}^{N-1} x(nT)\,Q(n,t) \tag{1.30}$$

where the interpolation functions Q(n,t) are chosen to compensate for the limited number of time samples available [5,6].

1.2.2 Numerical Approximations to the Fourier Transform

Rather than recovering the signal from its samples, there is often interest in calculating the Fourier transform of a signal from its samples. One might be interested in what frequencies are present in a speech sound or the critical resonances in an off-shore oil platform.

A straightforward way to approximate the integral involved in the Fourier transform (1.17) is to use the standard "rectangular rule" for numerical integration. In this method of numerical integration, the desired integral is approximated with a sum in a way which corresponds to convolving impulses located at the sampling instants, having weights equal to the sample values, with a rectangular pulse with a width W equal to the step size. With $s_n = s(nW)$

$$\overline{S}(\omega) = \int_{-\infty}^{\infty} \sum_{-\infty}^{\infty} s_n\, p(t-nW)\, e^{-j\omega t}\, dt \tag{1.31}$$

By interchanging the order of integration and summation and evaluating the transform of the rectangular pulse, p(t), (1.31) can be simplified, using (1.24), to give

$$\overline{S}(\omega) = \left[\sum_{n=-\infty}^{\infty} s_n\, e^{-j\omega nW} \right] \frac{W(\sin \omega W/2)}{(\omega W/2)} \tag{1.32}$$

The approximation to the Fourier transform integral in (1.32) is the product of two terms. The second term is independent of the actual values of the time samples, and is sometimes called a "diminishing factor" [7] since it decreases with increasing frequency. The first term, one of the main topics of this book, is called the Fourier transform of the discrete signal s_n (FTD).

There are many ways to approximate the Fourier integral, using other, more sophisticated numerical integration techniques [7]. Most of the approximations to the Fourier integral involve a calculation of the Fourier transform of a discrete-time signal (FTD) as in the rectangular rule approximation used in (1.31).

1.2.3 The Fourier Transform of a Discrete-Time Signal

The Fourier transform of the infinite-length discrete-time signal x_n is defined to be

$$X(\omega) = \sum_{n=-\infty}^{\infty} x_n\, e^{-j\omega n} \tag{1.33}$$

and the inverse transform is

$$x_n = 1/2\pi \int_{-\pi}^{\pi} X(\omega)\, e^{j\omega n}\, d\omega \tag{1.34}$$

This Fourier transform is defined for all values of the normalized frequency variable, ω. The transform has a continuous frequency variable, but a discrete time variable. In the review of the Fourier series (see Section 1.1.1), the same situation existed with a continuous time variable and a discrete frequency variable, so this mixture of continuous and discrete variables should not be too confusing. In fact, except for a change of variables, (1.33) is the same as (1.6), and (1.34) is the same as (1.5).

One of the main differences between the Fourier transform of a discrete-time signal (1.33) and the Fourier transform of a continuous-time signal (1.17) is the fact that the Fourier transform of a discrete-time signal is a periodic function of the frequency variable ω. For integer values of m,

$$X(\omega+2\pi m) = \sum_{n=-\infty}^{\infty} x_n\, e^{-j(\omega+2\pi m)n} = X(\omega) \tag{1.35}$$

This periodicity is present for exactly the same reason as the periodicity or aliasing discussed in connection with the sampling theorem (see Section 1.2.1). In (1.34), the discrete-time signal is being represented as a superposition of exponentials with various frequencies ω in the form of $e^{j\omega n}$.

An exponential with a frequency $\omega + 2\pi m$ for integer m will pass through exactly the same time samples as the exponential with frequency ω.

$$e^{j\omega n} = e^{j(\omega+2\pi m)n} \tag{1.36}$$

For this reason, the Fourier transform of a discrete-time signal *must* be a periodic function of frequency.

Example 1-4: Fourier Transform of a Discrete-Time Rectangular Pulse

A discrete-time rectangular pulse is defined as

$$p_n = \begin{cases} 1 & |n| \le W/2 \\ 0 & |n| > W/2 \end{cases} \tag{1.37}$$

as shown in Figure 1-7 for W = 4.

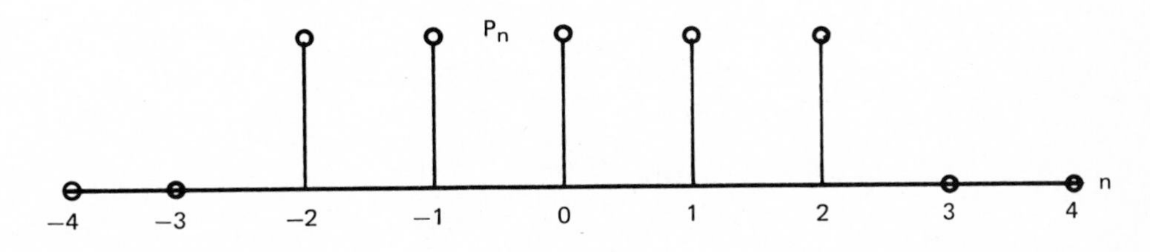

FIGURE 1-7. A DISCRETE-TIME RECTANGULAR PULSE

From (1.33),

$$P(\omega) = \sum_{-\infty}^{\infty} p_n e^{-j\omega n} = \sum_{-W/2}^{W/2} e^{-j\omega n} \tag{1.38}$$

This sum can be evaluated, using the following formula for the sum of a geometric series:

$$\sum_{n=M}^{n=N} a^n = \frac{a^M - a^{N+1}}{1 - a} \tag{1.39}$$

Evaluation of the sum in (1.38) using (1.39) yields

$$P(\omega) = \frac{\sin((W+1)\omega/2)}{\sin(\omega/2)} \tag{1.40}$$

This function, $P(\omega)$, appears often in discrete-time signal processing and is called the "Dirichlet kernel". It is shown in Figure 1-8.

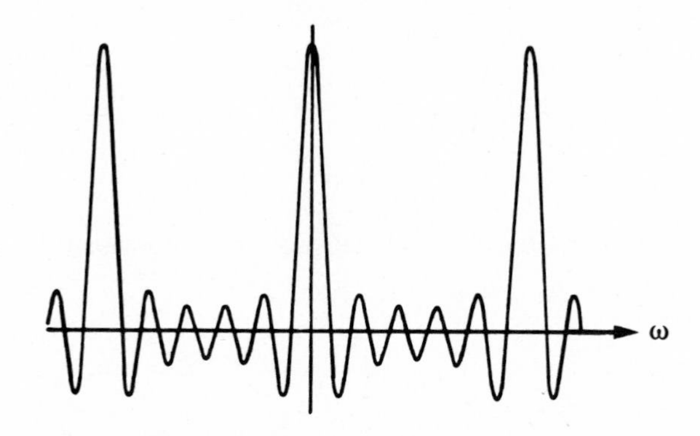

FIGURE 1-8. THE DIRICHLET FUNCTION

1.2.4 The Discrete Fourier Transform (DFT)

A discrete-time signal x_n, defined on the finite set of integers $0 \leq n \leq N-1$, can be expressed in a discrete version of the Fourier series as

$$x_n = 1/N \sum_{k=0}^{N-1} X_k e^{j(2\pi/N)kn} \tag{1.41}$$

where the coefficients X_k are given by the Discrete Fourier Transform (DFT) of x_n

$$X_k = \sum_{n=0}^{N-1} x_n e^{-j(2\pi/N)kn} \tag{1.42}$$

To emphasize the relation of the coefficients X_k to the signal x_n, (1.42) is often written as

$$X_k = \mathrm{DFT}\,(x_n) \tag{1.43}$$

Since the expansion is a periodic function of n with period N, (1.41) may be considered to be an expansion of the periodic signal $x_p(n)$ with period N shown as

$$x_p(n) = \sum_{m=-\infty}^{\infty} x_r\,(n-mN) \tag{1.44}$$

where

$$x_r(n) = \begin{cases} x_n & 0 \le n \le N-1 \\ 0 & \text{elsewhere} \end{cases} \tag{1.45}$$

Thus, (1.41) may be viewed in two ways [8]. When (1.41) is considered to represent a periodic signal, the coefficients X_k in (1.42) are sometimes called the *discrete Fourier series coefficients*. When (1.41) is viewed as an expansion of x_n on the finite set $0 \le n \le N-1$, (1.42) is called the *Discrete Fourier Transform (DFT)*.

Example 1-5: DFT of a Rectangular Pulse

In this example, the DFT of the length-8 signal, shown in Figure 1-9, is computed.

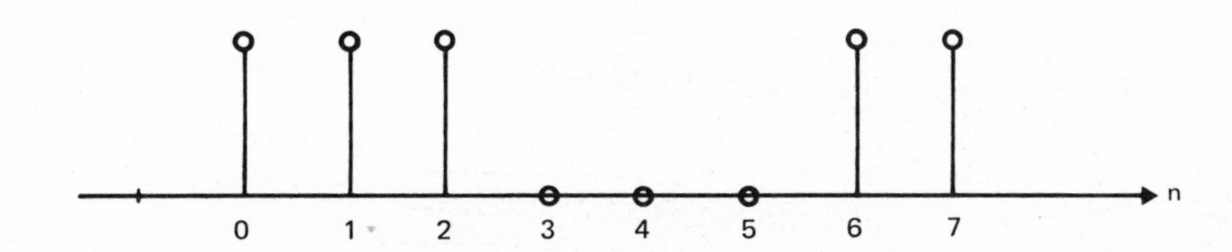

FIGURE 1-9. A LENGTH-8 SIGNAL

The signal

$$x(n) = \begin{cases} 1 \text{ for } n = 0,1,2,6,7 \\ 0 \text{ for } n = 3,4,5 \end{cases} \tag{1.46}$$

Using the definition of the DFT, (1.42), for N = 8:

$$X(k) = \sum_{n=0}^{7} x_n\, e^{-j\pi nk/4} \tag{1.47}$$

Inserting the values of x_n in (1.47) gives

$$X(k) = \sum_{n=0}^{2} e^{-j\pi nk/4} + \sum_{n=6}^{7} e^{-j\pi nk/4} \tag{1.48}$$

Since $e^{-j\pi nk/4}$ is periodic with a period of 8,

$$X(k) = \sum_{n=-2}^{2} e^{-j\pi nk/4} \tag{1.49}$$

By using the formula for the sum of a geometric series (1.39), this expression can be evaluated to give

$$X(k) = \frac{\sin(5k\pi/8)}{\sin(k\pi/8)} \tag{1.50}$$

1.2.5 The z Transform

The z transform of a discrete-time signal x_n is defined as

$$X(z) = \sum_{n=-\infty}^{\infty} x_n z^{-n} \tag{1.51}$$

The set of values of the complex variable z, for which the infinite sum in (1.51) converges, is called the region of convergence for X(z). For example,

$$x_n = \begin{cases} (1/2)^n & n \geq 0 \\ 0 & n < 0 \end{cases} \tag{1.52}$$

has a z transform

$$X(z) = \sum_{n=0}^{\infty} (1/2)^n z^{-n} = \frac{1}{1 - (1/2)z^{-1}} \tag{1.53}$$

This sum of a geometric series converges as long as $|(1/2)z^{-1}| < 1$. In other words, the region of convergence in the z plane is the set containing all z such that $|z| > 1/2$.

The inverse z transform is

$$x_n = 1/2\pi j \int_C X(z) z^{n-1} dz \tag{1.54}$$

where the path of integration C, a closed contour encircling the origin in the complex z plane, must be contained in the region of convergence for X(z).

When the region of convergence of the z transform includes the unit circle in the z plane, the Fourier transform of the discrete-time signal x_n is simply the z transform of x_n evaluated on the unit circle.

The z transform plays the same role in the analysis of discrete-time systems as the Laplace transform does for continuous-time systems. Just as the Laplace transform converts differential equations into algebraic equations, the z transform converts difference equations into algebraic equations. Once the z transform of a system output is obtained, the discrete-time output signal can be obtained with the inverse z transform (1.54), or with the help of tables of z transforms.

1.3 COMPARISON OF CONTINUOUS-TIME AND DISCRETE-TIME TRANSFORMS

Figure 1-10 shows a comparison of continuous-time and discrete-time transforms for frequency-domain representation of signals. Each transform for discrete-time signals is paired with its counterpart for continuous-time signals. Sampling in time leads to periodicity in frequency, while sampling of frequencies leads to periodicity in time.

1.3.1 Fourier Series - DFT

These are signal representations using discrete, harmonically related frequencies. The discrete frequency representation results in periodic time functions, both continuous-time and discrete-time. The spectrum of the discrete-time signal is periodic while the spectrum of the continuous-time signal is not.

1.3.2 Fourier Transforms

The Fourier transform of a continuous-time signal and the Fourier transform of a discrete-time signal are continuous-frequency representations of the corresponding signals. The transforms are functions of the real frequency variable ω. The spectrum of the discrete-time signal is periodic, while the spectrum of the continuous-time signal is not.

1.3.3 Laplace and z Transforms

These two transforms are functions of a complex variable. They are both integral transforms, requiring integration in the complex plane for the evaluation of the inverse transform. For the inverse Laplace transform, the Bromwich contour Br [3] is used, while for the inverse z transform, a closed contour C, encircling the origin of the z plane, is used. If the region of convergence for these transforms includes the frequency axis (imaginary axis in the s plane, unit circle in the z plane), then the Laplace and z transforms, when evaluated on the frequency axis, are equal to the Fourier transforms in Section 1.3.2.

CONTINUOUS-TIME TRANSFORMS

Fourier Series

$$x(t) = \sum_{k=-\infty}^{\infty} c_k\, e^{j2\pi kt/T}$$

$$c_k = \frac{1}{T} \int_{-T/2}^{T/2} x(t)\, e^{-j2\pi kt/T}\, dt$$

Fourier Transform

$$x(t) = 1/2\pi \int_{-\infty}^{\infty} X(\omega)\, e^{j\omega t}\, d\omega$$

$$X(\omega) = \int_{-\infty}^{\infty} x(t)\, e^{-j\omega t}\, dt$$

Laplace Transform

$$x(t) = 1/2\pi j \int_{Br} X(s)\, e^{st}\, ds$$

$$X(s) = \int_{-\infty}^{\infty} x(t)\, e^{-st}\, dt$$

DISCRETE-TIME TRANSFORMS

DFT

$$x(n) = 1/N \sum_{k=0}^{N-1} X_k\, e^{j2\pi kn/N}$$

$$X_k = \sum_{n=0}^{N-1} x(n)\, e^{-2\pi kn/N}$$

Fourier Transform (Discrete Signal)

$$x(n) = 1/2\pi \int_{0}^{2\pi} X(\omega)\, e^{j\omega n}\, d\omega$$

$$X(\omega) = \sum_{-\infty}^{\infty} x(n)\, e^{-j\omega n}$$

z Transform

$$x(n) = 1/2\pi j \int_{C} X(z)\, z^{n-1}\, dz$$

$$X(z) = \sum_{n=-\infty}^{\infty} x(n)\, z^{-n}$$

FIGURE 1-10. COMPARISON OF CONTINUOUS-TIME AND DISCRETE-TIME TRANSFORMS

1.4 CONVOLUTION AND LINEAR SYSTEMS

While Fourier techniques certainly aid signal analysis, they also play an important role in the study of linear time-invariant systems.

The time-domain description of the input-output characteristics of a linear time-invariant system with an input x(t), unit impulse response h(t), and output y(t) is given by the *convolution integral*

$$y(t) = \int_{-\infty}^{\infty} h(t-u)\, x(u)\, du \tag{1.55}$$

In (1.55), the superposition integral combines the response to unit impulses applied at various times u, h(t − u), weighted by the actual input value at time u, x(u), to get the output at time t.

The Fourier transform of the convolution in (1.55) is

$$Y(\omega) = H(\omega)\,X(\omega) \tag{1.56}$$

In other words, the convolution of two signals in the time domain corresponds to the product of the Fourier transforms of the two signals [3]. This property, as it applies to discrete-time signals, will be used in Chapter 3 to compute convolution using FFT algorithms for the required Fourier transforms.

The Fourier transform $H(\omega)$ of the unit impulse response $h(t)$ is known as the *frequency response* of the system. If the system input is a cosine of frequency ω

$$x(t) = \cos\omega t \tag{1.57}$$

then the system output may also be expressed as a cosine of the same frequency ω, with a phase shift and change in amplitude:

$$y(t) = |H(\omega)| \cos(\omega t + \theta) \tag{1.58}$$

The complex-valued frequency response has a magnitude $|H(\omega)|$ and a phase θ.

1.5 REFERENCES

1. Rabiner, L.R.,et.al., "Terminology in Digital Signal Processing" reprinted in SELECTED PAPERS IN DIGITAL SIGNAL PROCESSING,II. New York, NY: IEEE Press, 1976.

2. Stearns, S.D., DIGITAL SIGNAL ANALYSIS. Rochelle Park, NJ: Hayden Book Company, Inc., 1975.

3. Papoulis, A., THE FOURIER INTEGRAL AND ITS APPLICATIONS. New York, NY: McGraw-Hill Book Company, Inc., 1962.

4. Childers, D.G., Editor, MODERN SPECTRUM ANALYSIS. New York, NY: IEEE Press, 1978.

5. Jerri, A.J., "The Shannon Sampling Theorem - Its Various Extensions and Applications: A Tutorial Review," PROCEEDINGS OF THE IEEE, Vol. 65, No. 11, 1977, 1565-1596.

6. Oetken,G., Parks, T.W., and Schussler, H.W., "New Results in the Design of Digital Interpolators,"IEEE TRANS. ON ASSP, Vol.23, 1975, 301-309.

7. Achilles, D., "Pipeline Fourier Transform with Implicit Spline Interpolation," AEU, Vol. 29, 1975, 74-80.

8. Oppenheim, A.V., and Schafer, R.W., DIGITAL SIGNAL PROCESSING. Englewood Cliffs, NJ: Prentice-Hall, Inc., 1975.

Chapter 2

THE DISCRETE FOURIER TRANSFORM

The Discrete Fourier Transform (denoted the DFT), along with discrete convolution, is one of the two fundamental operations in digital signal processing. The DFT is used in the description, representation, and analysis of discrete-time signals. It is also used in conjunction with efficient algorithms for very rapid calculation of convolution and correlation. Indeed, the development of efficient DFT algorithms makes much of digital signal processing practical.

The understanding of the theory, algorithms, and programs of efficient methods for calculating the DFT on modern systems is one of the main goals of this book. This chapter presents the useful properties of the DFT, followed by various techniques to calculate the DFT.

2.1 PROPERTIES OF THE DISCRETE FOURIER TRANSFORM (DFT)

This section covers the main properties of the DFT which are useful for applications and algorithm development. Some of these are analogous to those of the continuous-time Fourier series and Fourier transforms reviewed in Sections 1.1 and 1.3, and some are strictly a discrete time result. Programming one of the DFT algorithms and experimenting with it should be informative and illustrate the points of this section. For other details and properties not covered here, excellent references are [1,2,3].

2.1.1 The Spectrum of a Signal

A discrete-time signal can be viewed as a finite length sequence of numbers. For example, x(n) having a length of seven could be represented as

$$x(n) = 3,\ -4,\ 39,\ 0,\ 1/3,\ \sqrt{7},\ -.065 \tag{2.1}$$

This sequence could represent a week of daily samples of some continuous-time process,or it could be an ordered list of parameters of seven objects having nothing to do with time. In most cases, the signal x(n) does come from samples of some continuous-time signal and, therefore, the index n is considered to be a sample of time.

The DFT of a signal describes its periodic character. Each component of the DFT is the amount of that signal that varies at some specific frequency. For example, let the signal x(n) be a set of samples of a pure sine wave having one frequency, shown by

$$x(n) = \cos(2\pi n/N) \tag{2.2}$$

The DFT of this signal is defined as

$$X(k) = \sum_{n=0}^{N-1} x(n)\, W^{nk} \tag{2.3}$$

where $W = e^{-j2\pi/N}$. The DFT becomes

$$X(k) = \begin{cases} 1/2\,N & \text{for } k = 1 \text{ and } N - 1 \\ 0 & \text{for all other } k \end{cases} \tag{2.4}$$

Only two elements of the DFT are nonzero, and they are the terms $X(1) = 1/2N$ and $X(N-1) = 1/2N$, which are the effects desired from a signal constructed from samples of one cycle of a sinusoid. If the samples had been of two cycles, $X(2)$ and $X(N-2)$ would have been the nonzero elements, and if samples of other multiples had been used, the corresponding frequencies would have been present.

If the signal is an even sequence of real numbers, the DFT always occurs in pairs with equal value. $X(k)$ always equals $X(N-k)$ because the trigonometric functions require two exponential terms for their definition. This is easily seen from Euler's relation:

$$\cos(x) = \frac{e^{jx} + e^{-jx}}{2}$$

The DFT of x(n) is a direct display of the frequency components of x(n) and is, therefore, called the spectrum of x(n) or the frequency-domain description of x(n).

A single pulse

$$d(n) = \begin{cases} 1 & \text{for } n = 0 \\ 0 & \text{for all other } n \end{cases} \tag{2.5}$$

has a DFT given by

$$\begin{aligned} X(k) &= \sum_{n=0}^{N-1} x(n)\, W^{nk} \\ &= \sum_{n=0}^{N-1} d(n)\, W^{nk} \\ &= 1 \quad \text{for all } k \end{aligned} \tag{2.6}$$

This means that a pulse has equal components at all frequencies. Other examples should be worked out to show that the DFT is a frequency description of a signal.

2.1.2 The Inverse DFT

Equation (2.3) shows how to calculate the DFT from a given signal, but an equally important problem is to find the signal from the DFT. The equation for doing this is very similar to that for the DFT. The inverse DFT is denoted the IDFT and has the property of

$$x(n) = \text{IDFT}\{X(k)\} = \text{IDFT}\{\text{DFT}\{x(n)\}\} \tag{2.7}$$

and is calculated by

$$x(n) = (1/N) \sum_{k=0}^{N-1} X(k)\, W^{-nk} \tag{2.8}$$

The validity of this equation is easily seen by substituting (2.8) into (2.3) and showing an identity of

$$\begin{aligned} X(k) &= \sum_{n=0}^{N-1} \left[(1/N) \sum_{i=0}^{N-1} X(i)\, W^{-ni} \right] W^{nk} \\ X(k) &= X(k) \end{aligned} \tag{2.9}$$

The only difference in the DFT and the IDFT is the factor of (1/N) and the negative exponent in the IDFT. In other words, if an algorithm exists for calculation of the DFT, multiplication of the DFT by 1/N and reversing the order of the elements give an algorithm for the IDFT. This symmetry proves to be important both in theory and application.

2.1.3 The Cyclic Convolution Property

A fundamental operation in linear digital signal processing is convolution, and the DFT is closely connected to it. This subject is developed more fully in Chapter 3 but is introduced here. Regular (also called noncyclic or aperiodic) convolution of two signals, x(n) and h(n), is defined by

$$y(n) = \sum_{m=-\infty}^{\infty} x(m)\, h(n-m) \tag{2.10}$$

and is denoted by

$$y(n) = x(n) * h(n)$$

If the sequence x(n) is of length N and h(n) is of length M, the length of the output sequence y(n) is N + M − 1. When used in signal processing, convolution is sometimes called a running average, where the h(n) are the weights in the average. The type of convolution that is associated with the DFT is similar, but the differences are important to understand.

Cyclic convolution is defined by a similar equation

$$y(n) = \sum_{m=0}^{N-1} x(m)\, h(n-m) \tag{2.11}$$

but in this case, all three sequences are the same length, which is N. Both x(n) and h(n) are assumed to be periodically extended outside the range 0 to N − 1, and this results in y(n) being periodic as well. For an alternate view of this periodic characteristic, consider the independent variables in (2.11) as all evaluated modulo N. This is denoted [8] by

$$\text{The residue of n modulo N} = \langle n \rangle_N$$

The definition of cyclic convolution can be written using this notation as

$$y(\langle n \rangle_N) = \sum_{m=0}^{N-1} x(\langle m \rangle_N)\, h(\langle n-m \rangle_N) \tag{2.12}$$

If the modulus is obvious from the context, the subscript is omitted.The convolution property of the DFT is given by the following relation:

$$\mathrm{DFT}\{y(n)\} = \mathrm{DFT}\{x(n)\}\,\mathrm{DFT}\{h(n)\} \tag{2.13}$$

This states that the DFT of the cyclic convolution of two signals is the product of the individual DFTs of the signals. The DFT converts cyclic convolution into multiplication. This property is used to reduce the amount of arithmetic needed for convolution by using the following equation:

$$y(n) = \mathrm{IDFT}\{\mathrm{DFT}\{x(n)\}\,\mathrm{DFT}\{h(n)\}\} \tag{2.14}$$

where some efficient algorithm is used for the DFT and IDFT.

Regular convolution is often what is required in signal processing applications; however, the convolution property holds for cyclic convolution which leads to a problem. The conversion of cyclic convolution to regular convolution is discussed in Chapter 3.

2.1.4 The Periodic Properties of x(n) and X(k)

Strictly speaking, x(n) is defined only over the range of n from 0 to N − 1. However, the IDFT of the DFT of x(n) can be calculated for values of n outside the original range. If that is done, it is found that the resulting x(n) is periodic with period N. In a similar manner, if X(k) is evaluated outside the 0 to N − 1 range, it too is found to be periodic. Mathematically, this is stated as

$$\begin{aligned} x(n) &= x(n+N) \qquad \text{for all } n \\ X(k) &= X(k+N) \qquad \text{for all } k \end{aligned} \tag{2.15}$$

This property may be expressed in terms of the signals being periodic, or also in terms of the indices being evaluated modulo N.

$$\begin{aligned} x(n) &= x(\langle n \rangle_N) \\ X(k) &= X(\langle k \rangle_N) \end{aligned} \tag{2.16}$$

The advantage of the first view is that it does not require number theory, while the advantage of the second view is that it generalizes to other problems, such as those in Section 2.2 and in Chapter 3.

2.1.5 Shifting and Modulation Properties

When the signal x(n) is shifted in time, the spectrum X(k) is multiplied by a linear phase shift. If

$$X(k) = \text{DFT}\{x(n)\} \tag{2.17}$$

then

$$\text{DFT}\{x(n+K)\} = X(k)\, e^{j(2\pi/N)Kk} \tag{2.18}$$

This states that a time shift of K in the signal sequence results in a linear phase shift in the spectrum of $(2\pi K/N)k$. Since the DFT and IDFT are essentially the same operation, a symmetry occurs. If the signal is multiplied by an exponential, the spectrum is shifted.

$$\text{DFT}\{x(n)\, e^{j(2\pi/N)Mn}\} = X(k-M) \tag{2.19}$$

The operation of multiplication by

$$e^{jx} = \cos x + j \sin x \tag{2.20}$$

is called modulation, and it shifts the spectrum of the modulated signal to the location of the modulating "carrier". Expressing e^{jx} in terms of sines and cosines is called Euler's relation [8].

2.1.6 The Sampling Property

The DFT of a subset of the original signal values can be found in terms of the DFT of the signal if the subset is viewed as a set of samples of the original signal. If X(k) is the DFT of x(n)

$$X(k) = DFT\{x(n)\} \tag{2.21}$$

and $x_s(n)$ is the sample sequence obtained from every Kth value of x(n)

$$x_s(n) = x(Kn) \qquad \text{for } n = 0, 1, 2, \ldots (N/K) - 1 \tag{2.22}$$

where N is a multiple of K (i.e., N = LK), the DFT of x_s becomes

$$DFT\{x_s(n)\} = (1/K) \sum_{m=0}^{K-1} X(k + mN/K) \tag{2.23}$$

This states that the DFT of the samples is formed from a shifted and summed version of the original DFT. In fact, this looks very much like the Fourier transform of the samples of a continuous-time signal in terms of the shifting and adding. The same sort of aliasing can occur if the shifted X(k) overlap.

2.1.7 Relation of the DFT to the z Transform

The z transform of a length-N sequence x(n) is denoted by

$$Z\{x(n)\} = X(z) \tag{2.24}$$

and defined by

$$X(z) = \sum_{n=0}^{N-1} x(n)\, z^{-n} \tag{2.25}$$

From this definition and equation (2.3), it is easily seen that the DFT of x(n) can be obtained by evaluating the z transform at values of

$$z = W^{-k} = e^{j2\pi k/N}$$

$$DFT\{x(n)\} = X(z)|_{z=W^{-k}} = X(W^{-k}) \tag{2.26}$$

In other words, the DFT of x(n) is the set of samples of X(z) at uniform spacing around the unit circle in the z plane.

2.1.8 Properties Resulting from Real Data and Certain Symmetries

From the definition of the DFT, certain properties can be seen for the case where the data has particular restrictions. The definition of X(k) as the DFT of x(n) is expanded by

$$\begin{aligned} X(k) &= \sum_{n=0}^{N-1} x(n)\, W^{nk} \\ &= \sum_{n=0}^{N-1} x(n)\, e^{-j2\pi nk/N} \\ &= \sum_{n=0}^{N-1} x(n)\, [\, \cos(2\pi/N)nk - j \sin(2\pi/N)nk \,] \end{aligned} \tag{2.27}$$

From this form, the real and imaginary parts of X(k) can be analyzed, and the following can be stated:

If x(n) is real, the real part of X(k) is an even function of k and the imaginary part is odd.

If x(n) is imaginary, the real part of X(k) is odd and the imaginary part is even.

In terms of these discrete-time signals, an even function has

$$x(n) = x(N-n) = x(-n)$$

and an odd function has

$$x(n) = -x(N-n) = -x(-n)$$

For the magnitude and phase of the spectrum, similar results can be stated:

If x(n) is real, the magnitude of X(k) is even and the phase is odd.

The results are summarized in Table 2-1.

TABLE 2-1. CHARACTERISTICS OF THE PARTS OF THE DFT

u(n) REAL PART OF x(n)	v(n) IMAGINARY PART x(n)	U(k) REAL PART OF X(k)	V(k) IMAGINARY PART X(k)
even	0	even	0
odd	0	0	odd
0	even	0	even
0	odd	odd	0

Table 2-1 can be used for combinations of restrictions. For example, if x(n) is real but neither even nor odd, its DFT will have a nonzero even real part and odd imaginary part as stated above. It shows that a general complex length-N x(n) has 2N degrees of freedom, a real x(n) has N degrees of freedom, and an even real x(n) has only N/2. These properties are used to improve the efficiency of certain special algorithms when used on restricted data (e.g., real data DFTs) [3,4].

2.2 CALCULATION OF THE DISCRETE FOURIER TRANSFORM (DFT)

In this section, several different approaches to the calculation of the DFT are presented and discussed. In particular, algorithms that improve calculation speed are emphasized. However, this proves very dependent on the hardware that is used to implement the algorithm. In the past, multiplications were very slow compared to other operations, so algorithms were developed to minimize the number of required multiplications at the expense of other operations. New hardware, such as the Texas Instruments TMS32010 digital signal processor, has very fast multiplication capability and may make such concerns unimportant. This book focuses on algorithms that take advantage of this feature.

The purpose of this section is to clearly show the theory and mathematics of several DFT algorithms and the efficient realization of them. The various tradeoffs in terms of the costs of program size, data transfers, indexing, etc., as well as arithmetic are discussed. To illustrate the algorithms more clearly than with a mathematical equation, FORTRAN subroutines are given and related to the constraints of possible realizations.

The structure of the development begins with the mathematics, goes to a flowgraph and/or FORTRAN program, and culminates in an assembly language program. The primary purpose of the programs is to illustrate the theories and principles of the various algorithms (not to be a collection of packages to be copied and run). The more detailed FORTRAN and assembly language programs are located in Chapters 4 and 5, respectively.

2.2.1 Basic Definitions and Constraints

If the data sequence x(n) is of length N, there are N terms in the DFT of x(n). There are practical applications where all N values of X(k) are not needed or where some of the data values are zero and, therefore, do not need to be taken into account. In developing efficient algorithms to calculate the DFT, it is important to know exactly what the problem requires. If the full number of frequency terms is L but only M of them are needed and there are N nonzero data terms, the DFT becomes

$$X(k) = \sum_{n=0}^{N-1} x(n)\, W^{nk} \tag{2.28}$$

but now

$$W = e^{-j2\pi/L} \text{ and k takes on M values}$$

and the number of complex multiplications required is NM.

In most cases, N = L and in some cases, M < L. In fewer cases, all three numbers are different. It is shown in the following sections that the most efficient algorithms for N = L = M are generally not the most efficient for the M << N case.

2.2.2 Direct Calculation of the DFT

An obvious way to calculate the DFT of a signal x(n) is to implement the definition of the DFT directly. This is done by computing the desired values of X(k) according to the formula in (2.28). In the case where x(n) is complex, the real and imaginary parts of x(n) are denoted by X(n) and Y(n), respectively, and the real and imaginary parts of the transform by A(k) and B(k). This is stated as

$$\begin{aligned} x(n) &= X(n) + j\,Y(n) \\ X(k) &= A(k) + j\,B(k) \end{aligned} \tag{2.29}$$

with X, Y, A and B being real-valued functions of integer variables. In terms of these real and imaginary parts, the DFT becomes

$$A(k) + j\,B(k) = \sum_{n=0}^{N-1} [X(n) + j\,Y(n)]\,[\cos(Qnk) - j\,\sin(Qnk)] \tag{2.30}$$

or

$$\begin{aligned} A(k) &= \sum_{n=0}^{N-1} X(n)\cos(Qnk) + Y(n)\sin(Qnk) \\ B(k) &= \sum_{n=0}^{N-1} X(n)\cos(Qnk) - Y(n)\sin(Qnk) \end{aligned} \tag{2.31}$$

where $Q = 2\pi/N$ and k take on any or all values between 0 and $N-1$. A simple FORTRAN program using X(n) and Y(n) for the real and imaginary parts of the data, and A(K) and B(K) for the real and imaginary parts of the DFT, is shown in Figure 2-1.

```
          Q = 6.2832/N
          DO 20 K = 1, N
             A(K) = 0
             B(K) = 0
             DO 10 J = 1, N
                A(K) = A(K) + X(J) * COS(Q * (J-1) * (K-1))
     +                      + Y(J) * SIN(Q * (J-1) * (K-1))
                B(K) = B(K) + Y(J) * COS(Q * (J-1) * (K-1))
     +                      - X(J) * SIN(Q * (J-1) * (K-1))
10           CONTINUE
20        CONTINUE
```

FIGURE 2-1. PROGRAM FOR DIRECT DFT CALCULATION

The amount of arithmetic required by an algorithn is an important measure of its efficiency. The calculation of all N values of A and of B from complex data requires $4N^2$ floating-point multiplications, $4N^2$ floating-point additions (and subtractions), and $4N^2$ trigonometric function evaluations plus numerous integer multiplications and additions.

In most mathematical expressions, the natural range of the indices is from zero to $N-1$, but in conventional FORTRAN, the indices cannot begin at zero and, therefore, go from one to N. Keep this in mind when relating the mathematical theory to the FORTRAN programs and when programming in conventional FORTRAN. Some modern high-level languages, such as Pascal, "C", or FORTRAN 77, allow indices starting at zero and are easier to use in programming mathematical formulas.

A modified program that halves the trigonometric function evaluations and considerably reduces the integer arithmetic is given in Figure 2-2 and in Chapter 4.

```
          Q = 6.2832/N
          DO 20 K = 1, N
             W = Q * (K-1)
             AT = X(1)
             BT = Y(1)
             DO 10 J = 2, N
                D = W * (J-1)
                C = COS(D)
                S = -SIN(D)
                AT = AT + C * X(J) - S * Y(J)
                BT = BT + C * Y(J) + S * X(J)
10           CONTINUE
             A(K) = AT
             B(K) = BT
20        CONTINUE
```

FIGURE 2-2. MODIFIED PROGRAM FOR A DIRECT DFT CALCULATION

This program is still inefficient in that it evaluates the sine and cosine functions $2N^2$ times while only N different values are needed (only N/4 if N is even). One solution consists of some form of table lookup from a precomputed table of sines and cosines. The program in Figure 2-3 is an example of that approach.

```
          Q = 6.2832/N
          DO 5 K = 1, N
             C(K) = COS(Q * (K-1))
             S(K) = -SIN(Q * (K-1))
5         CONTINUE
          DO 20 K = 1, N
             AT = X(1)
             BT = Y(1)
             I =K
             DO 10 J = 2, N
                AT = AT + C(I) * X(J) - S(I) * Y(J)
                BT = BT + C(I) * Y(J) + S(I) * X(J)
                I = I + K -1
                IF (I.GT.N) I = I -N
10           CONTINUE
             A(K) = AT
             B(K) = BT
20        CONTINUE
```

FIGURE 2-3. DFT CALCULATION WITH TRIG TABLE LOOKUP

In this program, there is a problem with indexing the sines and cosines. In order to utilize a table of reasonable size and with little redundancy, the index must be evaluated modulo N. That operation is in the statement just preceding the label 10 and requires a test and branching that takes time, especially if it is located in the innermost loop. Further reduction of the table size could be made but only at the expense of complicating the indexing.

In the above programs, a rather small amount of arithmetic is carried out in the inside loop (the DO 10 loop). If the cost of testing and branching done in each loop is appreciable compared to the cost of doing the arithmetic in the loop, an improvement in speed can be made at the expense of longer program code. In fact, if N were fairly small, the inside loop could be removed, and the five statements repeated N times with appropriate indices in so-called straight-line code. This turns out to be desirable in many modern systems with high-speed arithmetic.

It is unnecessary to use pure straight-line code, but a modified form can be used that increases the calculation inside the innermost loop until the time for testing and branching is negligible compared to the other required time. The next program is an example of that modification.

```
          Q = 6.283185307179586/N
          DO 20 K = 1, N
             W = Q * (K-1)
             AT = 0
             BT = 0
             J = 0
10           CONTINUE
                D = W * (J)
                C = COS(D)
                S = -SIN(D)
                J = J +1
                AT = AT + C * X(J) - S * Y(J)
                BT = BT + C * Y(J) + S * X(J)
                D = W * (J)
                C = COS(D)
                S = -SIN(D)
                J = J +1
                AT = AT + C * X(J) - S * Y(J)
                BT = BT + C * Y(J) + S * X(J)
             IF (J.LT.N) GO TO 10
             A(K) = AT
             B(K) = BT
20        CONTINUE
```

FIGURE 2-4. PROGRAM USING PARTIAL STRAIGHT-LINE CODE

An increase in code length is seen when compared to the program in Figure 2-2. For this approach, the length of the DFT must be a multiple of the block length in the loop. Remember that each of these program examples illustrates a principle but is not optimized to any particular situation.

The characteristics of the specific hardware must be taken into account when evaluating any algorithm. Tradeoffs in the amount of data multiplications and additions, the complexity of the indexing, the use of registers and memory, and the amount of testing and branching must be made. Some of these points are discussed by Morris [5].

2.2.3 The Goertzel Algorithm for the DFT

The next approach is rather different from the various forms of direct calculation of the DFT. In Section 2.1.7, it was shown that the DFT is the z transform of x(n) evaluated on the unit circle. In other words, calculating the DFT is the same as polynomial evaluation. The modified z transform of x(n) is given by

$$X(z) = \sum_{n=0}^{N-1} x(n)\, z^n \tag{2.32}$$

which for clarity in this development uses a positive exponent unlike the usual definition of equation (2.25). This is illustrated for a length-4 sequence, which is a third-order polynomial, by

$$X(z) = x(3)z^3 + x(2)z^2 + x(1)z + x(0) \tag{2.33}$$

The DFT from (2.28) can be found by evaluating (2.32) at $z = W^k$, which can be written as

$$X(k) = X(z)\big|_{z=W^k} = \text{DFT}\{x(n)\} \tag{2.34}$$

where

$$W = e^{-j2\pi/N}$$

The most efficient way of evaluating a polynomial is by "Horner's rule" [7], also called nested evaluation. This is illustrated for the polynomial in (2.33) by

$$X(z) = ((x(3)\,z + x(2))\,z + x(1))\,z + x(0) \tag{2.35}$$

This sequence of operations can be written as a linear difference equation in the form of

$$y(m) = z\,y(m-1) + x(N-m) \tag{2.36}$$

with initial condition $y(-1) = 0$ and the desired result being the solution at $m = N$. This is given by

$$X(z) = y(N) \tag{2.37}$$

This can be viewed as a first-order filter with the input being the data sequence in reverse order and the value of the polynomial at z being the solution sampled at $m = N$. Applying this to the DFT gives the Goertzel algorithm [1] which is

$$y(m) = W^k\,y(m-1) + x(N-m), \qquad y(-1) = 0 \tag{2.38}$$

with

$$X(k) = y(N)$$

The flowgraph of the algorithm is shown in Figure 2-5, and a simple FORTRAN program is given in Figure 2-6.

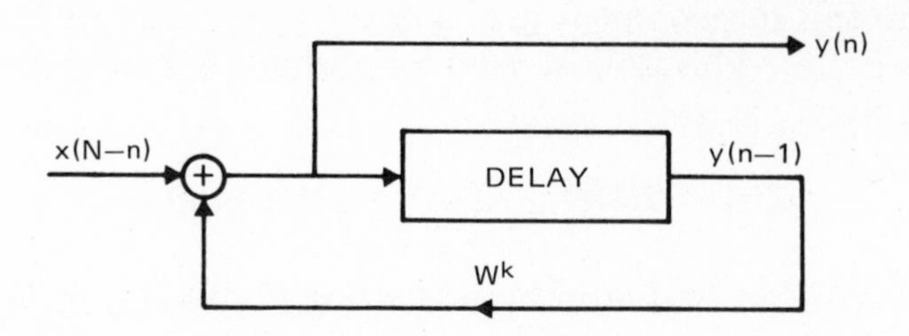

FIGURE 2-5. THE GOERTZEL ALGORITHM AS A RECURSIVE FILTER

```
      Q = 6.2832/N
      DO 20 K = 1, N
         S = SIN(Q * (K−1))
         C = COS(Q * (K−1))
         AT = 0
         BT = 0
         DO 10 J = 1, N
            T = C * AT + S * BT + X(N−J+1)
            BT = S * AT − C * BT + Y(N−J+1)
            AT = T
10       CONTINUE
         A(K) = AT
         B(K) = BT
20    CONTINUE
```

FIGURE 2-6. PROGRAM OF A FIRST-ORDER GOERTZEL ALGORITHM

When comparing this program with the direct calculation in Figure 2-2, it is seen that the number of floating-point multiplications and additions are the same. In fact, the structures of the two algorithms look similar, but close examination shows that the way the sine and cosine enter the calculations is very different. In Figure 2-2, new sine and cosine values are calculated for each frequency and for each data value, while for the Goertzel algorithm in Figure 2-6, they are calculated only for each frequency in the outer loop. Because of the recursive or feedback nature of the algorithm, the sine and cosine values are "updated" each loop rather than recalculated. This results in 2N trigonometric evaluations rather than $2N^2$. It also results in an increase in quantization error.

One of the reasons the Goertzel algorithm did not improve efficiency is that the constant in the feedback or recursive path is complex and, therefore, requires four real multiplications. A modification of the scheme to make it second order removes the complex multiplications and reduces the number of required multiplications by two. Using Euler's relation of

$$W^k = e^{-j2\pi k/N} = \cos(2\pi k/N) - j\sin(2\pi k/N) \tag{2.39}$$

with (2.35), and using subscripts rather than functional notation give

$$\begin{aligned} y_m &= W^k\, y_{m-1} + x_{N-m} \\ &= [\cos - j\sin]\, y_{m-1} + x_{N-m} \end{aligned} \tag{2.40}$$

$$\text{and} \qquad y_{m-1} = [\cos - j\sin]\, y_{m-2} + x_{N-m+1} \tag{2.41}$$

Multiplying both sides of (2.41) by $-$ [cos + j sin] and adding to (2.40) give

$$y_m = [2\cos]\, y_{m-1} - y_{m-2} + x_{N-m} - [\cos + j\sin]\, x_{N+1-m} \tag{2.42}$$

The recursive part now has all real multiplications (and one is by unity), but the input data now has a complex coefficient.

Further algebraic manipulation gives

$$q_m = [2\cos]\, q_{m-1} - q_{m-2} + x_{N-m} \tag{2.43}$$

and

$$y_m = q_m - [\cos + j\sin]\, q_{m-1} \tag{2.44}$$

Equation (2.43) is a form of state equation with very simple coefficients. In fact, only one multiplication per iteration is required. The complication has been moved to the output equation (2.44), but it is evaluated only once after m reaches N. This is shown by

$$X(k) = y_N = q_N - [\cos + j\sin]\, q_{N-1} \tag{2.45}$$

with (2.43) having initial conditions of

$$q_0 = q_{-1} = 0 \tag{2.46}$$

Still more algebra allows the data x(n) to input the difference equation in a forward or ascending order rather than the reverse order of (2.43). Equations (2.43) and (2.44) become

$$q_m = [2\cos]\, q_{m-1} - q_{m-2} + x_m \tag{2.47}$$

and

$$y_m = q_m - [\cos - j\sin]\, q_{m-1}$$

A FORTRAN implementation of these two equations is given in Figure 2-7 and in Chapter 4.

```
         Q = 6.2832/N
         DO 20 K = 1, N
            S = SIN(Q * (K-1))
            C = COS(Q * (K-1))
            CC = 2 * C
            A2 = 0
            A1 = X(1)
            B2 = 0
            B1 = Y(1)
            DO 10 J = 2, N
               T = A1
               A1 = CC * A1 - A2 + X(J)
               A2 = T
               T = B1
               B1 = CC * B1 - B2 + Y(J)
               B2 = T
10          CONTINUE
            A(K) = C * A1 - A2 - S * B1
            B(K) = S * A1 + C * B1 - B2
20       CONTINUE
```

FIGURE 2-7. PROGRAM FOR THE SECOND-ORDER GOERTZEL DFT ALGORITHM

In relating this program to the difference equation (2.47), several points should be noted. In the difference equation, the time index goes from 0 to $N-1$, but traditional FORTRAN has indices starting from one. Therefore, the program index J goes from 1 to N. Since the first iteration is trivial, its effect is put in the initial condition and the iterations are started at $J = 2$. When the program index J reaches N, the difference equation time index is $N-1$ so that one more iteration must be calculated but with no input. In the program of Figure 2-7, this last iteration is included with the calculations of the output equation (2.47) in the two lines in the program between labels 10 and 20. The required arithmetic is approximately halved if the data is real.

When compared to the direct DFT or the first-order Goertzel algorithm, this program uses half the multiplications, the same number of additions, and 2N trigonometric evaluations which could be stored in a precomputed table (if space is available).

2.2.4 The Chirp z-Transform Method

In the previous section, the DFT was viewed as an evaluation of the z-transform polynomial on the unit circle using Horner's rule. This resulted in Goertzel's algorithm which calculated each DFT value separately by sampling the output of a recursive filter. Now another filtering approach is used–one very different from the Goertzel algorithm. Recall the definition

$$X(k) = \sum_{n=0}^{N-1} x(n)\, W^{nk} \qquad (2.48)$$

The identity

$$(k - n)^2 = k^2 - 2kn + n^2 \tag{2.49}$$

$$nk = \frac{1}{2}[n^2 - (k - n)^2 + k^2]$$

is used to rearrange the indices of the DFT giving

$$X(k) = \left\{ \sum_{n=0}^{N-1} [x(n)\, W^{n^2/2}]\, W^{-(k-n)^2/2} \right\} W^{k^2/2} \tag{2.50}$$

This can be interpreted as first multiplying (modulating) the data by a chirp sequence, then convolving (filtering) it, and then finally multiplying the output by a chirp to give the DFT. If a chirp signal is defined as

$$h(n) = W^{n^2/2} \tag{2.51}$$

the chirp transform in (2.50) becomes

$$X(n) = \{[x(n)\, h(n)] * h^{-1}(n)\}\, h(n) \tag{2.52}$$

This approach is very versatile in that it not only can calculate the DFT as presented here, but with small modifications, it can evaluate the z transform on certain contours in the z plane other than the unit circle. It can also concentrate the evaluated points on the contours rather than have them uniformly distributed as calculated by the FFT (discussed in Section 2.2.8).

If implemented on digital hardware, the chirp z transform does not seem advantageous for calculating the normal DFT. Therefore, no programs are presented here. For unusual applications where it might have some advantages, see [2,6] for discussion and [4] for a program in FORTRAN.

2.2.5 Rader's Conversion of the DFT into Convolution

Another conversion of the DFT into convolution is developed in this section. In contrast to the chirp z transform, no pre- or post-multiplications are required, and the convolution is cyclic. This method requires use of some number theory, which can be found in a very readable form in [8] and is easy enough to verify on one's own.

The two basic operations in digital signal processing are the DFT in (2.53) and cyclic convolution in (2.54).

$$X(k) = \sum_{n=0}^{N-1} x(n)\, W^{nk} \tag{2.53}$$

$$y(k) = \sum_{n=0}^{N-1} x(n)\, h(k\text{-}n) \tag{2.54}$$

In both cases, the indices are evaluated modulo N. If it is desired to convert the DFT of (2.53) into the convolution of (2.54), the nk product must be changed to the $k-n$ difference, or more generally a sum. With normal real numbers, this can be done with logarithms, but it is more complicated when working in a finite set of integers modulo N. From number theory [8,9], it can be shown that if the modulus is a prime number, a base (called a primitive root) exists such that a form of integer logarithm can be defined. This is stated in the following way. If N is a prime number, a number r exists such that the integer equation

$$n = r^m \bmod N \tag{2.55}$$

or

$$n = \langle r^m \rangle_N$$

creates a unique, one-to-one map of the $N-1$ member set $m = \{0,\ldots,N-2\}$ and the $N-1$ member set $n = \{1,\ldots,N-1\}$. This is illustrated for $N = 5$ in Table 2-2.

TABLE 2-2. THE INTEGERS r^m MODULO 5

r	m=0	1	2	3	4	5	6	7	...
1	1	1	1	1	1	1	1	1	
2	1	2	4	3	1	2	4	3	
3	1	3	4	2	1	3	4	2	
4	1	4	1	4	1	4	1	4	
5	*	0	0	0	*	0	0	0	
6	1	1	1	1	1	1	1	1	

* not defined

From this table, it is easy to see that there are two primitive roots, 2 and 3, and equation (2.55) defines a permutation of the integers n from the integers m (except for zero). Equation (2.55) and a primitive root (usually chosen to be the smallest of those that exist) can be used to convert the DFT in (2.53) to the convolution in (2.54).

Since (2.55) cannot give a zero, a new length-(N-1) data sequence is defined from x(n) by removing the term with index zero. Let

$$n = r^{-m}$$

and

$$k = r^s$$

where the term with the negative exponent is defined as the integer that

$$\langle r^{-m} r^{m} \rangle_N = 1$$

If N is a prime number, r^{-m} always exists. For example, $\langle 2^{-1} \rangle_N = 3$. Equation (2.53) now becomes

$$X(r^s) = \sum_{m=0}^{N-2} x(r^{-m}) W^{r^{-m} r^{s}} + x(0), \qquad s=0,1,..,N-2 \tag{2.56}$$

and

$$X(0) = \sum_{n=0}^{N-1} x(n)$$

New functions are defined, which are simply a permutation in order of the original functions, as

$$x'(m) = x(r^{-m}), \qquad X'(s) = X(r^s), \qquad W'(n) = W^{r^n} \tag{2.57}$$

Equation (2.56) becomes

$$X'(s) = \sum_{m=0}^{N-2} x'(m) W'(s-m) + x(0) \tag{2.58}$$

which is cyclic convolution of length N − 1 (plus x(0)) denoted as

$$X'(k) = x'(k) * W'(k) + x(0) \tag{2.59}$$

Applying this change of variables (use of logarithms) to the DFT can best be illustrated from the matrix formulation of the DFT. Equation (2.53) is written as

$$\begin{bmatrix} X(0) \\ X(1) \\ X(2) \\ X(3) \\ X(4) \end{bmatrix} = \begin{bmatrix} 0 & 0 & 0 & 0 & 0 \\ 0 & 1 & 2 & 3 & 4 \\ 0 & 2 & 4 & 1 & 3 \\ 0 & 3 & 1 & 4 & 2 \\ 0 & 4 & 3 & 2 & 1 \end{bmatrix} \begin{bmatrix} x(0) \\ x(1) \\ x(2) \\ x(3) \\ x(4) \end{bmatrix} \tag{2.60}$$

where the square matrix should be the terms of W^{nk}. For the sake of clarity, only the exponents nk are shown. Removing the x(0) term, applying the mapping of (2.57), and using r = 2 (and $r^{-1} = 3$) give

$$\begin{bmatrix} X(1) \\ X(2) \\ X(4) \\ X(3) \end{bmatrix} = \begin{bmatrix} 1 & 3 & 4 & 2 \\ 2 & 1 & 3 & 4 \\ 4 & 2 & 1 & 3 \\ 3 & 4 & 2 & 1 \end{bmatrix} \begin{bmatrix} x(1) \\ x(3) \\ x(4) \\ x(2) \end{bmatrix} + \begin{bmatrix} x(0) \\ x(0) \\ x(0) \\ x(0) \end{bmatrix} \tag{2.61}$$

and

$$X(0) = x(0) + x(1) + x(2) + x(3) + x(4)$$

which can be seen to be a reordering of the structure in (2.60). This is in the form of cyclic convolution as indicated in (2.58).

Rader showed this in 1968 [8], stating that a prime length-N DFT could be converted into a length-(N − 1) cyclic convolution of a permutation of the data with a permutation of the W's. He also stated that a more complicated version of the same idea would work for a DFT with a length equal to a prime to a power. The details of that theory can be found in [8].

Until 1975, this conversion approach received little attention since it seemed to offer few advantages. For example, a length-17 DFT could be converted into a length-16 convolution which could be performed by an FFT, but this would require more work than it would save. The real advantage of this scheme emerged when new fast cyclic convolution algorithms were developed by Winograd.

2.2.6 Winograd's Prime-Length DFTs

In 1975, S. Winograd developed a theory for efficient calculation of prime-length cyclic convolution using a minimum number of multiplications. This approach is fairly complicated, therefore, the details will not be developed here. They can be found in [8] and [9] and a brief description is given in Section 3.2.3. The results are important to the topic of this chapter because the optimal convolution algorithms can be combined with Rader's conversion of the DFT to convolution to give very efficient short, prime-length DFT algorithms. This theory was a breakthrough and resulted in excellent practical algorithms for lengths of 3, 5, and 7 which can be found in the short DFT modules in the PFA program in Chapter 4 and in [8,9,12,13]. Compromise versions [10] gave good results for lengths of 11, 13, 17, and 19. Winograd's algorithm for calculating longer transforms with composite lengths is known as the WFTA and is covered in Section 2.2.13.

2.2.7 Multidimensional Index Mapping

One of the most practical methods of reducing the arithmetic necessary to calculate the DFT is to use an index mapping (a change of variables) in such a way as to change a one-dimensional DFT into a two- or higher dimensional DFT. In fact, this is the main idea behind the very efficient Cooley-Tukey and Winograd algorithms. The goal of index mapping is to change a single difficult problem into multiple easy ones.

This is possible because of the special structure of the DFT.

In the case where the length of the DFT is not prime, N can be factored as $N = N_1N_2$ with the time index n of (2.53) taking on values of

$$n = 0, 1, 2, \ldots, N-1 \tag{2.62}$$

Two new independent variables are defined by

$$\begin{aligned} n_1 &= 0, 1, 2, \ldots, N_1-1 \\ n_2 &= 0, 1, 2, \ldots, N_2-1 \end{aligned} \tag{2.63}$$

and the completely general linear equation which maps n_1 and n_2 to n is given by

$$n = \langle K_1n_1 + K_2n_2 \rangle_N \tag{2.64}$$

This defines a relation between all allowed n_1 and n_2 in (2.63) and a value for n. The question is whether all of the n in (2.62) are represented, i.e., whether the map is unique (one-to-one). It has been shown [11] that certain K_i always exist such that the map in (2.64) is unique. Two cases are considered as follows:

The notation for the greatest common divisor [8,9] of two numbers, M and N, used here is

$$(N,M) = L$$

where both N and M are some multiple of L.

Case 1: N_1 and N_2 are relatively prime, i.e., $(N_1,N_2) = 1$

The integer map of (2.64) is unique if and only if the following:

$$(K_1 = aN_2) \text{ and/or } (K_2 = bN_1) \text{ and } (K_1,N_1) = (K_2,N_2) = 1 \tag{2.65}$$

Case 2: N_1 and N_2 are not relatively prime, i.e., $(N_1,N_2) \rangle 1$.

The integer map of (2.64) is unique if and only if the following:

$$\begin{aligned} &(K_1 = aN_2) \text{ and } (K_2 \neq bN_1) \text{ and } (a,N_1) = (K_2,N_2) = 1 \\ \text{or} \\ &(K_1 \neq aN_2) \text{ and } (K_2 = bN_1) \text{ and } (K_1,N_1) = (b,N_2) = 1 \end{aligned} \tag{2.66}$$

Reference [11] should be consulted for the details of these conditions and examples.

Two classes of index maps are defined from these conditions. The map of (2.64), which is given again by

$$n = \langle K_1 n_1 + K_2 n_2 \rangle_N$$

is called a prime factor map (PFM) when

$$\text{PFM:} \qquad K_1 = aN_2 \quad \text{and} \quad K_2 = bN_1 \tag{2.67}$$

The map of (2.64) is called a common factor map (CFM) when

$$\text{CFM:} \qquad K_1 = aN_2 \quad \text{or} \quad K_2 = bN_1, \qquad \text{but not both.} \tag{2.68}$$

This can be a bit confusing because of the choice “and/or” in the conditions of (2.65). The PFM can be used only if the factors are relatively prime, but the CFM can be used whether they are relatively prime or not, i.e., it can always be used.

The PFM was used by Good and by Thomas in their DFT algorithms. Cooley and Tukey used the CFM in their DFT algorithms, some having common factors ($N = 2^M$) and some not (mixed radix) [8].

A map similar to (2.64) is defined for the frequency index as

$$k = \langle K_3 k_1 + K_4 k_2 \rangle_N \tag{2.69}$$

and the same conditions, (2.65) and (2.66), hold for the uniqueness of this map in terms of K_3 and K_4.

Next, these index maps are applied to the definition of the DFT by defining the two-dimensional arrays for the input data and its DFT as

$$\begin{aligned} \hat{x}(n_1,n_2) &= x(K_1 n_1 + K_2 n_2) \\ \hat{X}(k_1,k_2) &= X(K_3 k_1 + K_4 k_2) \end{aligned} \tag{2.70}$$

The substitution of these changes of variables into the definition of the DFT given in (2.3) or (2.53) results in

$$\hat{X} = \sum_{n_2=0}^{N_2-1} \sum_{n_1=0}^{N_1-1} \hat{x}\, W^{K_1K_3n_1k_1}\, W^{K_1K_4n_1k_2}\, W^{K_2K_3n_2k_1}\, W^{K_2K_4n_2k_2} \tag{2.71}$$

Although this looks rather complicated, careful examination shows the amount of arithmetic required to be the same as in the direct calculation of (2.53). However, the constants K_i can be chosen in such a way that the calculations are “uncoupled” and the arithmetic is reduced. The requirements for this are

$$\langle K_1 K_4 \rangle_N \quad \text{and/or} \quad \langle K_2 K_3 \rangle_N = 0 \tag{2.72}$$

When this requirement and those for uniqueness in (2.64) are applied, it is found that the K_i may always be chosen such that one of the terms in (2.72) is zero, and if the N_i are relatively prime, it is possible to make both terms zero.

If the N_i are not relatively prime, only one of the terms may be set to zero. If they are relatively prime, it is possible to either set one or two to zero. There is a choice in that case. This in turn causes one or both of the center two W terms in (2.71) to become unity. The details are in [11].

An example of the CFM used in the Cooley-Tukey radix-4 FFT for a length-16 DFT is

$$N = 16 = 4^2$$

$$n = 4n_1 + n_2 \qquad k = k_1 + 4k_2 \tag{2.73}$$

The residue reduction in (2.64) is not needed here since n does not exceed N. The resulting arrays of indices are

(2.74)

n_2 \ n_1 =	0	1	2	3
0	0	4	8	12
1	1	5	9	13
2	2	6	10	14
3	3	7	11	15

= n

k_2 \ k_1 =	0	1	2	3
0	0	1	2	3
1	4	5	6	7
2	8	9	10	11
3	12	13	14	15

= k

Since (for this example) the factors of N have a common factor, only one of the conditions in (2.72) can hold and, therefore, (2.71) becomes

$$\hat{X} = \sum_{n_2=0}^{3} \sum_{n_1=0}^{3} \hat{x}\, W_4^{n_1k_1}\, W_{16}^{n_2k_1}\, W_4^{n_2k_2} \tag{2.75}$$

where the subscript for W is the denominator of the exponent which is also the period of W.

$$W_N = e^{-j2\pi/N}$$

This has the form of a two-dimensional DFT with an extra term W_{16}, called a "twiddle factor" [1]. The inner sum over n_1 represents four length-4 DFTs, the W_{16} term represents 16 complex multiplications, and the outer sum over n_2 represents another four length-4 DFTs.

This choice of the K_i "uncouples" the calculations since the first sum over n_i for $n_2 = 0$ calculates the DFT of the first row of the data array $x(n_1,n_2)$, and those data values are never needed in the succeeding calculations. The row calculations are independent, and examination of the outer sum shows that the column calculations are likewise independent. This is illustrated in Figure 2-8 for a length-15 DFT.

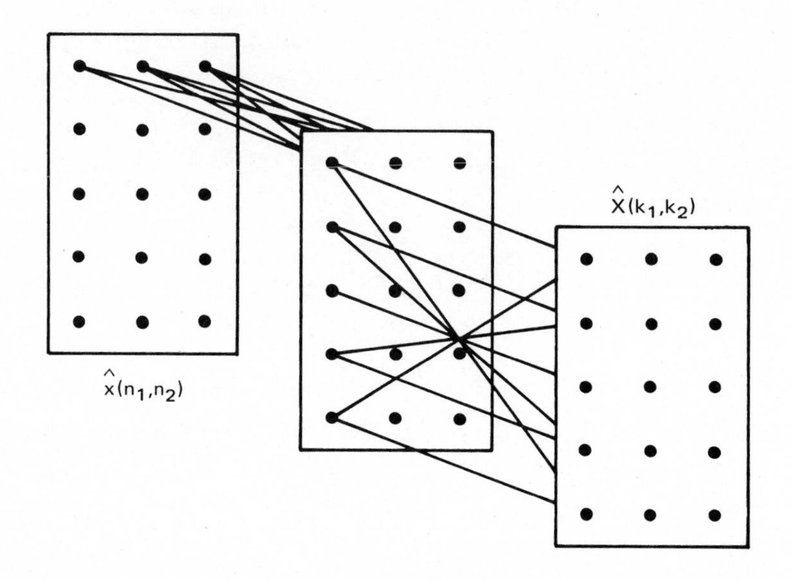

FIGURE 2-8. UNCOUPLING OF THE ROW AND COLUMN CALCULATIONS

The left 3-by-5 array is the mapped input data, the center array has the rows transformed, and the right array is the DFT array. The row DFTs and the column DFTs are independent of each other. If twiddle factors (TF) exist as in the center W in (2.75), that multiplication takes place on the center array of Figure 2-8.

This uncoupling feature reduces the amount of arithmetic required and allows the results of each row DFT to be written back over the input data locations, since that input row will not be needed again. This is called in-place calculation and results in a large memory requirement savings (discussed in the next section, Section 2.2.8).

An example of the CFM used when the factors of N are relatively prime is for N = 15 and given as

$$n = 15 = 3 \cdot 5$$

$$n = 5n_1 + n_2 \qquad (2.76)$$
$$k = k_1 + 3k_2$$

Again, the residue reduction is not explicitly needed. The resulting arrays of indices are

(2.77)

n_2 \ n_1 =	0	1	2
0	0	5	10
1	1	6	11
2	2	7	12
3	3	8	13
4	4	9	14

$= n$

k_2 \ k_1 =	0	1	2
0	0	1	2
1	3	4	5
2	6	7	8
3	9	10	11
4	12	13	14

$= k$

Although the factors 3 and 5 are relatively prime, use of the CFM sets only one of the terms in (2.72) to zero. The DFT in (2.71) becomes

$$\hat{X} = \sum_{n_2=0}^{4} \sum_{n_1=0}^{2} \hat{x}\, W_3^{n_1 k_1}\, W_{15}^{n_2 k_1}\, W_5^{n_2 k_2} \tag{2.78}$$

which has the same form as (2.75), including the existence of the twiddle factors (TF). Here the inner sum is five length-3 DFTs. This is shown in Figure 2-9 as uncoupled row and column calculations where each of the "blocks" are DFTs. Figures 2-8 and 2-9 show the same phenomenon, but in a slightly different way.

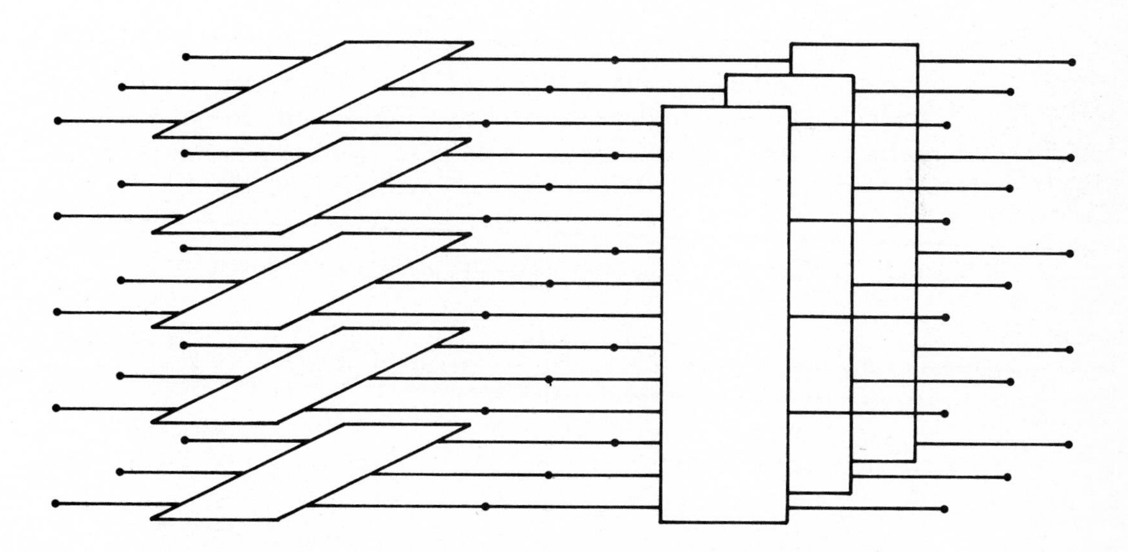

FIGURE 2-9. ROW DFTS FOLLOWED BY COLUMN DFTS

The PFM is next used with the same sequence, N = 15. This time, the situation of (2.65) with the "and" condition is used as (2.67) to give an index map of

$$\begin{aligned} n &= \langle 5n_1 + 3n_2 \rangle_{15} \\ k &= \langle 10k_1 + 6k_2 \rangle_{15} \end{aligned} \tag{2.79}$$

Here the residue reduction is necessary, and the resulting arrays of indices are

$$
\begin{array}{c|ccc} n_1 = & 0 & 1 & 2 \\ \hline 0 & 0 & 5 & 10 \\ 1 & 3 & 8 & 13 \\ n_2 = \; 2 & 6 & 11 & 1 \\ 3 & 9 & 14 & 4 \\ 4 & 12 & 2 & 7 \end{array} = n
\qquad
\begin{array}{c|ccc} k_1 = & 0 & 1 & 2 \\ \hline 0 & 0 & 10 & 5 \\ 1 & 6 & 1 & 11 \\ k_2 = \; 2 & 12 & 7 & 2 \\ 3 & 3 & 13 & 8 \\ 4 & 9 & 4 & 14 \end{array} = k
\tag{2.80}
$$

Since the factors of N are relatively prime and the PFM is being used, both terms in (2.72) are zero, and (2.71) becomes

$$\hat{X} = \sum_{n_2=0}^{4} \sum_{n_1=0}^{2} \hat{x}\, W_3^{n_1 k_1}\, W_5^{n_2 k_2}$$

which is the same as (2.78), except that now the PFM gives a pure two-dimensional DFT calculation with no TFs, and the sums can be interchanged, i.e., done in either order. This calculation is also illustrated in Figure 2-9.

Decomposition by index mapping is used to improve the arithmetic efficiency. Table 2-3 gives the number of complex multiplications necessary to calculate the DFT directly by (2.31) and (2.53), and compares it to calculation by the two-dimensional forms of (2.75), (2.78), and (2.81). Using the CFM in (2.75) requires four length-4 DFTs for the inner sum, and these DFTs require 4^2 multiplications each.

TABLE 2-3. NUMBER OF COMPLEX MULTIPLICATIONS FOR THE DFT

LENGTH	METHOD	NUMBER OF MULTIPLICATIONS
16	Direct	$16^2 = 256$
16	CFM of (2.75)	$4 \cdot 4^2 + 16 + 4 \cdot 4^2 = 144$
15	Direct	$15^2 = 225$
15	CFM of (2.78)	$5 \cdot 3^2 + 15 + 3 \cdot 5^2 = 135$
15	PFM of (2.81)	$5 \cdot 3^2 + 3 \cdot 5^2 = 120$

Practical algorithms, presented in Sections 2.2.9 and 2.2.10, use methods that have fewer than N^2 multiplications for the short DFTs. For example, length-4 DFTs require no multiplications and, therefore, for the length-16 DFT, only the TFs must be calculated. That calculation uses 16 multiplications, many fewer than the 256 or 144 indicated in Table 2-3.

2.2.8 In-Place Calculations and the Unscrambling Problem

The use of either the CFM or the PFM for simplifying the DFT allows the intermediate results to be written back over the original data in such a way that, after all the calculations are finished, the DFT will be in the array originally occupied by the input data sequence. Separate input and output arrays are then not required, thus allowing a substantial savings of memory [1,3].

Unfortunately, the use of in-place calculations causes the order of the DFT terms to be permuted, i.e., scrambled. This is seen by tracing the calculations through one of the examples of Section 2.2.7. In the CFM of (2.76) which results in the DFT of (2.78), the first sum is for $n_2 = 0$, and that sum over n_1 for $k_1 = 0$, 1, and 2 is a length-3 DFT which can be stored in the array locations formerly occupied by the data for $n_1 = 0$, 1, and 2. The same is done for $n_2 = 1$, 2, and 3. The array which originally was the input data is now filled with the row DFTs. Next, the 15 "twiddle-factor" (TF) multiplications are performed, and that result is also stored back in the original array. Finally, the outer sum over n_2 is calculated for each value of k_1. These column DFTs are stored back over the intermediate results. This completes the calculation of the DFT by the formula of (2.78), but the final DFT results are in locations specified by the input map of (2.76) whereas they should be in the location specified by the output map. It is helpful to refer to Figure 2-8 and 2-9 for illustration of these ideas.

For a radix-2 Cooley-Tukey decimation-in-frequency FFT, the input index map is set to

$$n = 8n_1 + 4n_2 + 2n_3 + n_4 \tag{2.81}$$

which, to uncouple the calculations, requires an output map of

$$k = k_1 + 2k_2 + 4k_3 + 8k_4 \tag{2.82}$$

After the DFT is calculated by in-place methods, the output values are in locations defined by (2.81) rather than (2.82). For certain applications, this scrambled output order is not important, but for most applications, the order must be unscrambled before the DFT can be considered complete. For the radix-2 FFT map, this is straightforward. Because the radix of the FFT is the same as the base for the binary number representation, the correct address for any term is found by reversing the binary bits. This is seen from the coefficients of the maps in (2.81) and (2.82). The part of most FFT programs that does this reordering is called a bit-reversed counter, or for other radices, a digit-reversed counter. Examples of unscrambling are in the FORTRAN programs in Chapter 4 and in references [1,2,3].

If the output map of (2.82) is used to derive the FFT algorithm so the correct output order can be obtained, the input must be reordered (scrambled) so that its values are in locations specified by the output map rather than the input map. This scrambling is the same digit-reversed counting as before, and the resulting algorithm is called a decimation-in-time FFT because of the order of the calculations [1,2,3].

The same process of a post-unscrambling or pre-scrambling occurs for the in-place calculations with the PFM. The details of this process are covered in [12].

2.2.9 The Cooley-Tukey FFT Algorithm

This section investigates the particular properties that result from using the CFM to calculate the DFT in an efficient way. The original algorithm of Cooley and Tukey [6,8] used this index map and obtained such impressive efficiencies that it truly revolutionized digital signal processing. It is informative to read the brief history of the FFT that is reprinted in [6,8]. The reading of Section 2.2.7 is necessary in order to properly understand this section on the Cooley-Tukey FFT.

The Cooley-Tukey FFT, called simply the FFT in this section, uses the CFM on both the time and frequency index of (2.64) to give

$$n = \langle K_1 n_1 + K_2 n_2 \rangle_N$$
$$k = \langle K_3 k_1 + K_4 k_2 \rangle_N$$

with conditions (2.68) which become

$$K_1 = aN_2 \quad \text{or} \quad K_2 = bN_1 \qquad \text{but not both}$$

and (2.83)

$$K_3 = cN_2 \quad \text{or} \quad K_4 = dN_1 \qquad \text{but not both}$$

Conditions for the uncoupling of the row and column calculations in (2.72) are given by

$$\langle K_1 K_4 \rangle_N = 0 \quad \text{or} \quad \langle K_2 K_3 \rangle_N = 0 \qquad \text{but not both} \tag{2.84}$$

In order that each short sum be a short DFT, the following must also hold:

$$\langle K_1 K_3 \rangle_N = N_2 \quad \text{and} \quad \langle K_2 K_4 \rangle_N = N_1$$

The simplest set of coefficients that satisfy all these conditions are

$$a = d = K_2 = K_3 = 1 \tag{2.85}$$

This gives for the index maps (2.82)

$$\begin{aligned} n &= N_2 n_1 + n_2 \\ k &= k_1 + N_1 k_2 \end{aligned} \tag{2.86}$$

These index maps are all evaluated modulo N as indicated in (2.64), but in (2.86), explicit reduction is not necessary since n does not exceed N. The reduction notation will be omitted for clarity. From (2.71) and example (2.75), the DFT is

$$\hat{X} = \sum_{n_2=0}^{N_2-1} \sum_{n_1=0}^{N_1-1} \hat{x}\, W_{N_1}^{n_1k_1} W_{N}^{n_2k_1} W_{N_2}^{n_2k_2} \tag{2.87}$$

This map of (2.86) and the form of the DFT in (2.87) are the fundamentals of the FFT. The independent row and column operations are illustrated in Figures 2-8 and 2-9.

Unlike the PFM, the order of the sums using the CFM in (2.87) cannot be reversed. This is because of the TF term. When the individual sums and their effects are considered separately, the first sum over n_1 is

$$f(k_1,n_2) = \sum_{n_1=0}^{N_1-1} \hat{x}(n_1,n_2)\, W_{N_1}^{n_1k_1} \tag{2.88}$$

This is carried out by first setting $n_2 = 0$, then calculating a length-N_1 DFT which is done by computing the sum in (2.88) N_1 times for each of the values of k_1. Next, n_2 is set to one, and the N_1 sums calculated again. This is done for each of the N_2 values of n_2. These are the horizontal calculations in Figure 2-8.

The original data array $x(n_1,n_2)$ was transformed along each of its rows to give $f(k_1,n_2)$, which is a function of frequency along its first dimension and still a function of time along its second dimension. This mixed function is exactly of the form necessary to multiply by the twiddle-factor (TF) map, the next step in carrying out (2.87) to give

$$g(k_1,n_2) = f(k_1,n_2)\, W_N^{k_1n_2} \tag{2.89}$$

This is simply N point-by-point multiplications of the two arrays.

The final step in calculating (2.87) is the outer sum over n_2 which represents DFTs of the columns of (2.89). This may be performed by stepping through the N_1 values of k_1 and calculating the DFTs. These are sums for each of the N_2 values of k_2 and are the column calculations in Figures 2-8 and 2-9. The calculation of the DFT is completed with

$$\hat{X}(k_1,k_2) = \sum_{n_2=0}^{N_2-1} g(k_1,n_2)\, W_{N_2}^{n_2k_2} \tag{2.90}$$

Another possibility for the outer sum is to further factor N_2 and to carry out N_1 two-dimensional operations as was done with the inner sum and the TF multiplication. This is how the FFT is usually programmed for more than two factors, and is illustrated in Figures 2-10 and11.

The pictorial description of the two-dimensional FFT in Figures 2-8 and 2-9 cannot be extended to higher dimensions. An alternative form that does not emphasize the multidimensions but extends to any dimension is shown in Figure 2-10. A flowgraph is used to show the operations on each element of the original data, x(n). This form is similar to that in [1,2,3]. The relation between Figures 2-9 and 2-10 can be seen in each stage of the flowgraph as working on a vector which is made up of a concatenation of the columns of the arrays in Figure 2-9. Four length-2 DFTs are followed by two length-4 DFTs. Note the scrambling of the output order.

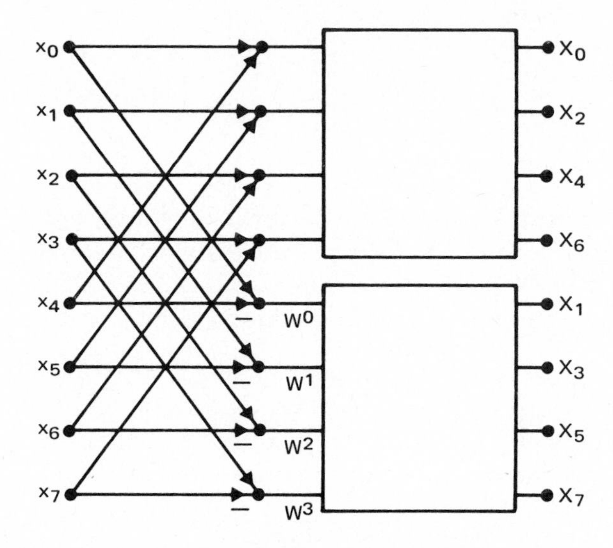

FIGURE 2-10. A TWO-FACTOR LENGTH-8 FFT DIAGRAM

2.2.10 The Decimation-in-Frequency FFT

The most common and most efficient form of the FFT uses all dimensions of the same length. This length is called the radix of the algorithm. The DFT of length N is thus related to the radix R by

$$N = R^M$$

which gives M dimensions, each of length R. Because the short length-R DFTs for $R = 2$ and 4 require no multiplications, and those for 8 and 16 require very few, radices of 2 and 4 or, occasionally other powers of two, are the most common.

For a radix-2 FFT, equation (2.86) becomes

$$n = (N/2)n_1 + n_2$$
$$k = k_1 + 2\,k_2$$

and (2.88) becomes

$$f(k_1,n_2) = \sum_{n_1=0}^{1} \hat{x}(n_1,n_2)\,(-1)^{n_1 k_1} \tag{2.91}$$

which is illustrated in (2.93). Equation (2.90) becomes

$$\hat{X}(k_1,k_2) = \sum_{n_2=0}^{N/2-1} g(k_1,n_2)\,W_{N/2}^{n_2 k_2} \tag{2.92}$$

These equations and Figure 2-10 show how a length-N DFT can be calculated in terms of two length-N/2 DFTs. This is the basic idea behind the radix-2 FFT. It is called a decimation-in-frequency (DIF) FFT because the two half-length DFTs give samples of the original DFT. One gives the even terms and the other the odd terms. This process is repeated on the two half-length DFTs to give the complete algorithm, which is illustrated in Figure 2-11.

The flowgraph for the length-2 DFT of (2.91) is

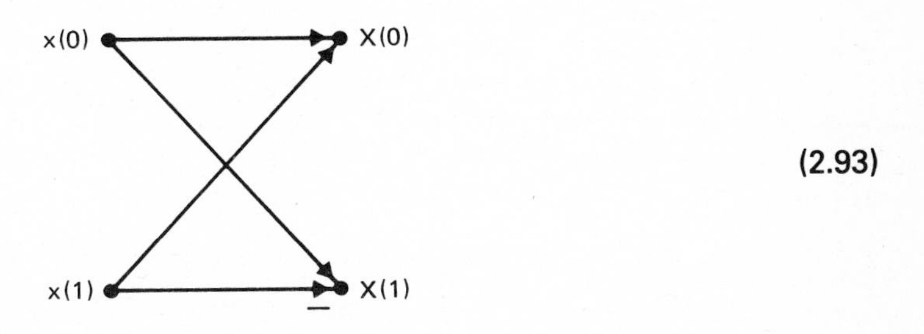

(2.93)

and it is called, because of its shape, a butterfly (BF). The FORTRAN statements (using FORTRAN indices) for this length-2 butterfly are

```
T    = X(1) + X(2)
X(2) = X(1) - X(2)
X(1) = T
T    = Y(1) + Y(2)
Y(2) = Y(1) - Y(2)
Y(1) = T
```

(2.94)

which calculate the DFT of a length-2 input with the real part in the array X and the imaginary part in the array Y and store the result back in the locations of the data. This is called "in-place" calculation and was discussed in Section 2.2.8. The in-place calculation does require one temporary storage location, which is called the variable T in (2.94). In (2.94), the FORTRAN indices are used rather than the time indices.

The flowgraph notation for a butterfly will follow that used by Rabiner and Gold [2] where the horizontal lines in (2.93) are omitted for clarity. The butterfly of (2.92) and (2.94) will be denoted by

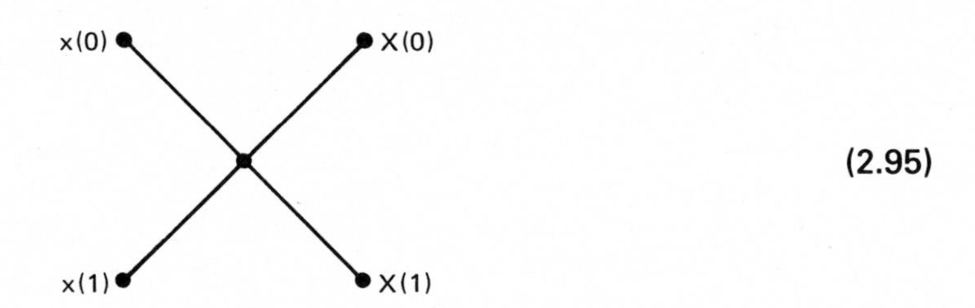

(2.95)

which stands for the mathematical operations given in (2.92). The flowgraph for N = 8 is given in Figure 2-11 using the butterfly notation of (2.95).

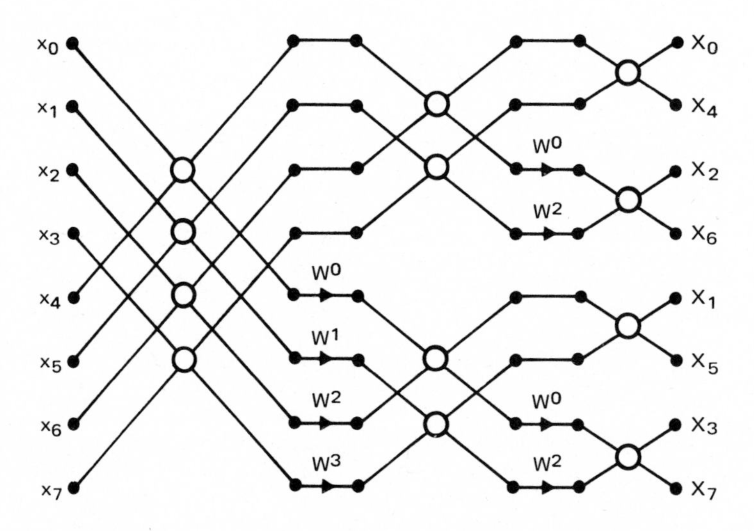

FIGURE 2-11. A FLOWGRAPH OF A LENGTH-8 FFT

This flowgraph is an extension of Figure 2-10 and should be related to the representations in Figures 2-8 and 2-9, and to the mathematical formula in (2.87). The graph has three "stages" corresponding to the three dimensions necessary in a radix-2 FFT for $N = 2^3 = 8$. The first stage is the sum over n_1 in (2.87) so that the intermediate variables are the terms in (2.88), which are multiplied by the TFs. Each column of that two-dimensional array (the middle array of Figure 2-9) is the upper and lower half of the output of the first stage. Those two length-4 sequences are each treated the same way as the original length-8 sequence; this gives the second stage. After the second multiplication by TFs, the last stage cannot be further factored and so is also calculated by butterflies.

All of the butterfly calculations are identical and, therefore, can be implemented by a single set of program instructions, such as (2.94) located in a loop. The TF terms are different and should be examined. The TF array in (2.87) and the first stage of Figure 2-10 is illustrated by the following array of the exponents of W:

$$\text{TF:} \qquad W_8^{k_1 n_2} = \begin{bmatrix} W^0 & W^0 \\ W^0 & W^1 \\ W^0 & W^2 \\ W^0 & W^3 \end{bmatrix} = \begin{bmatrix} 1 & 1 \\ 1 & W \\ 1 & -j \\ 1 & -jW \end{bmatrix} \tag{2.96}$$

The TF array will always have ones in the first column and first row for any radix. Therefore, the number of complex multiplications necessary to multiply by the TF array is two, not eight, as might at first be expected. The TF array after the second stage is

$$\text{TF:} \qquad W_4^{k_1 n_2} = \begin{bmatrix} 1 & 1 \\ 1 & -j \end{bmatrix} \tag{2.97}$$

which is used twice, but requires no actual multiplications, only a real and imaginary part exchange.

In most implementations of the FFT, the TF multiplications are not carried out in a separate operation but are absorbed into the butterfly and considered to be part of it.

2.2.11 The FFT Program Structure

Over the years, a rather standard nested loop structure has evolved for programming the FFT. The FORTRAN program, given in Figure 2-12, is an implementation of the ideas presented up to this point.

```
      N2 = N
      DO 10 K = 1, M
         N1 = N2
         N2 = N2/2
         E = 6.283185307179586/N1
         A = 0
         DO 20 J = 1, N2
            C = COS (A)
            S = -SIN (A)
            A = J * E
            DO 30 I = J, N, N1
               L = I + N2
               XT = X(I) - X(L)
               X(I) = X(I) + X(L)
               YT = Y(I) - Y(L)
               Y(I) = Y(I) + Y(L)
               X(L) = XT * C - YT * S
               Y(L) = XT * S + YT * C
30          CONTINUE
20       CONTINUE
10    CONTINUE
```

FIGURE 2-12. A RADIX-2 COOLEY-TUKEY FFT PROGRAM

The outer loop (the DO 10 loop) steps through the M dimensions of (2.90), carrying out (2.88) and (2.89) M times. It also implements the M = 3 stages of the flowgraph of Figure 2-11.

The next loop (the DO 20 loop) steps through the N_2 length-2 DFTs and the corresponding TFs (or butterflies). The inner loop (the DO 30 loop) steps through the columns.

Inside the DO 30 loop, the first statement is an address offset that is part of the index map. The next four statements are the length-2 DFT, or butterfly, similar to (2.94), and the last two are the TF multiplication.

This program is, in fact, an efficient FFT, very similar to many being used in practice. It is easy to count the number of multiplications for this type of algorithm. There are $M = \log_2 N$ stages, with each stage having approximately N/2 multiplies. This gives the familiar formula for the number of complex multiplications for an FFT as

$$\text{Multiplications} = (1/2)\, N \log_2 N \tag{2.98}$$

There are various tricks that can be used to reduce the required arithmetic even further, but they will only lower the multiplicative constant of (2.98). The basic N log N formula is a property of the staged or factored formulation of the FFT algorithm.

Soon after the original FFT was discovered, improvements were found. The first and probably the most significant means of reducing arithmetic was the use of other radices. The length-4 DFT requires no multiplications and, therefore, is an efficient radix. A set of FORTRAN statements to calculate a length-4 DFT in place is given by

```
R1 = X(1) + X(3)
R2 = X(1) - X(3)
R3 = X(2) + X(4)
R4 = X(2) - X(4)
S1 = Y(1) + Y(3)
S2 = Y(1) - Y(3)
S3 = Y(2) + Y(4)
S4 = Y(2) - Y(4)                    (2.99)
X(1) = R1 + R3
X(3) = R1 - R3
X(2) = R2 + S4
X(4) = R2 - S4
Y(1) = S1 + S3
Y(3) = S1 - S3
Y(2) = S2 - R4
Y(4) = S2 + R4
```

where the X array contains the real part of the data and the Y array contains the imaginary part of the data. These statements can easily be seen as a length-4 version of Figure 2-12.

The number of multiplications for an FFT with a radix of 2 or 4 is exactly the number of TF multiplications, N of which take place after each of the M stages. Since there are twice as many stages in a radix-2 algorithm as a radix-4, there should be twice as many multiplications in a radix-2 FFT as in a radix-4. Because some of the multiplications are by one and are not performed, the reduction in multiplications in going to a radix of 4 is not quite by a factor of two; however, this reduction is significant and is given later in Table 2-4. Not only the number of multiplications but also the number of additions is reduced. Since fewer stages are required, the amount of data transfer is also reduced.

A radix-4 FFT program is easily obtained by substituting the length-4 butterfly of (2.99) for the length-2 butterfly in Figure 2-12 and making a few other modifications. Several radix-4 FFTs can be found in Chapter 4.

The length-8 DFT can be developed to require only four real multiplications which, although not zero as for lengths of 2 and 4, are smaller than those required by the TFs. Unfortunately, the reduction in the number of stages is not as great as a factor of two, and the reduction in the number of additions is small or nonexistent.

Increasing the radix to 16 theoretically looks somewhat promising, but the decrease in multiplications is often offset by very long program code and an increase in additions.

The relative reductions in arithmetic possible with radices of 2, 4, 8, and 16 are given later in Table 2-4. Other radices are seldom used since they generally require substantially more multiplications for the butterfly and do not allow any reductions in the TF multiplications.

The second means of reducing arithmetic is to remove the unnecessary TF multiplications by plus or minus one or by plus or minus the square root of minus one (+/ −1 or +/−j). This occurs when the exponent of W_N is zero or a multiple of N/4. A reduction in the number of additions also occurs under these conditions.

In a program, this reduction is usually achieved by having a special butterfly for the TF that is one or j, or that has the condition shown in (2.101). A radix-2 program that has two butterflies and removes the multiplications by one is shown in Figure 2-13. Programs that have a radix of 2, 4, and 8 as well as those that remove multiplications by +/−1, +/−j, and reduce those by $+/-1/\sqrt{2}$ $+/-$ $j\sqrt{2}$ are listed and described in Chapter 4.

```
      N2 = N
      DO 10 K = 1, M
         N1 = N2
         N2 = N2/2
         DO 1 I = 1, N, N1
            L = I + N2
            XT = X(I) - X(L)
            X(I) = X(I) + X(L)
            YT = Y(I) - Y(L)
            Y(I) = Y(I) + Y(L)
            X(L) = XT
            Y(L) = YT
1        CONTINUE
         IF (K.EQ.M) GO TO 10
         E = 6.283185307179586/N1
         A = E
         DO 20 J = 2, N2
            C = COS (A)
            S = -SIN (A)
            A = J * E
            DO 30 I = J, N, N1
               L = I + N2
               XT = X(I) - X(L)
               X(I) = X(I) + X(L)
               YT = Y(I) - Y(L)
               Y(I) = Y(I) + Y(L)
               X(L) = XT * C - YT * S
               Y(L) = XT * S + YT * C
30          CONTINUE
20       CONTINUE
10    CONTINUE
```

FIGURE 2-13. A REDUCED MULTIPLY RADIX-2 FFT PROGRAM

This program should be compared with the single butterfly version in Figure 2-12 to see how the butterflies that require no multiplies are trapped.

When the exponent of W_N is N/8 or some multiple, a small reduction of multiplications is also possible. Since the real and imaginary parts of W to the N/8 power are equal, the following can be done: Let the real part of W be C, the imaginary part S, the real part of the data be X, and the imaginary part Y. The product of W times the data is then

$$
\begin{aligned}
(C + jS)(X + jY) &= R + jI \\
R &= CX - SY \\
I &= CY + SX
\end{aligned}
\tag{2.100}
$$

which requires four real multiplications and two real additions. If W has an exponent of N/8 or some multiple, C and S are equal to one over the square root of two, i.e., $C = S = 1/\sqrt{2}$. In that case, the product (2.100) becomes

$$(C + jC)(X + jY) = R + jI \tag{2.101}$$

$$
\begin{aligned}
R &= C(X - Y) \\
I &= C(X + Y)
\end{aligned}
\tag{2.102}
$$

which requires two real multiplications and two real additions.

Removing or reducing the extra arithmetic that seems to be required when the TFs have an exponent of a multiple of N/8 can improve execution speed in some cases, but this is an approach that must be evaluated carefully. In some hardware such as the TMS32010, the multiplications and additions are coupled to each other so that a reduction in one may not increase speed. In addition, tests to find out when the special conditions exist may take more time than the reduction in arithmetic saves. The use of special butterflies to reduce arithmetic also reduces quantization errors [18].

A third means of reducing multiplications does so at the expense of increasing additions. Normally, the complex multiplications in the FFT programs use the form of (2.100) which require four real multiplications and two real additions. An alternative formulation of complex multiplication requires three real multiplications and five real additions. Using the notation of (2.100) and defining three intermediate variables by

$$Z = C(X - Y) \tag{2.103}$$

$$
\begin{aligned}
D &= C + S \\
E &= C - S
\end{aligned}
\tag{2.104}
$$

the real and imaginary parts of the product can be written as

$$\begin{aligned} R &= DY + Z \\ I &= EX - Z \end{aligned} \tag{2.105}$$

If a form of table lookup is used for the TFs, D and E can be precalculated and stored in a table with C. The execution of the FFT would then require three real multiplications and three real additions for each complex multiplication. If multiplications are much slower than additions, this scheme will be faster. Morris has used this algorithm in his FFTs [5].

A time-consuming and unnecessary part of the execution of the FFT programs discussed so far is the generation of the sine and cosine terms which are the real and imaginary parts of W. There are basically three approaches to obtaining these sines and cosines. The first is to generate or calculate them as needed, as shown in the programs in Figures 2-12 and 2-13. The second is to generate some sine and cosine values and then update them as needed in a way similar to Goertzel's algorithm in Section 2.2.3. The fastest way is to precalculate the sine and cosine values and fetch them from a table as needed. A table lookup program is given in Figure 2-14. Examples of all these approaches are presented in Chapter 4.

```
C----------------------INITIALIZE SINE AND COSINE TABLES ----------
            P = 6.283185307179586/N
            DO 1 K = 1, N/2
               A = (K-1) * P
               WR(K) = COS(A)
               WI(K) = -SIN(A)
    1       CONTINUE
C----------------------RADIX-2 FFT------------------------------
            N2 = N
            DO 10 K=1, M
               N1 = N2
               N2 = N2/2
               IE = N/N1
               IA = 1
               DO 20 J=1, N2
                  C = WR(IA)
                  S = WI(IA)
                  IA = IA + IE
                  DO 30 I=J, N, N1
                     L = I + N2
                     XT = X(I) - X(L)
                     X(I) = X(I) + X(L)
                     YT = Y(I) - Y(L)
                     Y(I) = Y(I) + Y(L)
                     X(L) = XT * C - YT * S
                     Y(L) = XT * S + YT * C
   30             CONTINUE
   20          CONTINUE
   10       CONTINUE
```

FIGURE 2-14. A RADIX-2 FFT PROGRAM USING TABLE LOOKUP

In the program of Figure 2-14, a table of sine and cosine values is calculated by the first part. The second part calculates the FFT by fetching the TFs from the WR and WI arrays.

There are several other possible approaches to further improvement in execution speed and in versatility. A wider variety of data lengths is possible if other dimension lengths are mixed in with those considered here. Because these other lengths are generally less efficient, this results in an overall loss of efficiency, but the larger variety of lengths may make it desirable anyway. Singleton's program [4,6] is an excellent example of a highly developed mixed-radix Cooley-Tukey FFT algorithm.

If all the DFT values are not needed or if many of the data values are zero, advantage can be taken by a process called "pruning" which removes all possible unnecessary calculations. This is described by Markel in [6].

It should be remembered from Section 2.2.9 that in-place calculations result in a scrambling of the order of the output of an FFT program. An in-place unscrambler is presented as part of the FFT program in Chapter 4.

Index maps in addition to the form in (2.86), the resulting flowgraph of Figure 2-11, and the programs in Figures 2-12 and 2-13 can be used. The approach discussed so far is often called decimation-in-frequency because of the type of sampling of the DFT, shown in Figure 2-10, that the index map causes. An alternative which uses a scrambler followed by an FFT uses the output map rather than the input map. This results in what is called the decimation-in-time algorithm. These are covered in [1,2,3]; however, they are basically the same as the versions presented here.

To make some initial evaluations and to better understand some of the modifications made on the original radix-2 FFT, a table of the necessary multiplications and additions is presented in Table 2-4. These numbers reflect the real data arithmetic only and do not take into account indexing or addressing arithmetic.

TABLE 2-4. NUMBER OF REAL MULTIPLIES AND ADDS FOR DIFFERENT COMPLEX FFT ALGORITHMS

M	N	M1	M2	M3	M5	A1	A2	A3	M1+A1	M3+A3
RADIX = 2										
1	2	4	0	0	0	6	4	4	10	4
2	4	16	4	0	0	24	18	16	40	16
3	8	48	20	8	4	72	58	52	120	60
4	16	128	68	40	28	192	162	148	320	188
5	32	320	196	136	108	480	418	388	800	524
6	64	768	516	392	332	1152	1026	964	1920	1356
7	128	1792	1284	1032	908	2688	2434	2308	4480	3340
8	256	4096	3076	2568	2316	6144	5634	5380	10240	7948
9	512	9216	7172	6152	5644	13824	12802	12292	23040	18444
10	1024	20480	16388	14344	13324	30720	28674	27652	51200	41996
11	2048	45056	36868	32776	30732	67584	63490	61444	112640	94220
12	4096	98304	81924	73736	69644	147456	139266	135172	245760	208908
RADIX = 4										
1	4	12	0	0	0	22	16	16	34	16
2	16	96	36	28	24	176	146	144	272	172
3	64	576	324	284	264	1056	930	920	1632	1204
4	256	3072	2052	1884	1800	5632	5122	5080	8704	6964
5	1024	15360	11268	10588	10248	28160	26114	25944	43520	36532
6	4096	73728	57348	54620	53256	135168	126978	126296	208896	180916
RADIX = 8										
1	8	32	4	4	4	66	52	52	98	56
2	64	512	260	252	248	1056	930	928	1568	1180
3	512	6144	4100	4028	3992	12672	11650	11632	18816	15660
4	4096	65536	49156	48572	48280	135168	126978	126832	200704	175404
RADIX = 16										
1	16	80	20	20	20	178	148	148	258	168
2	256	2560	1540	1532	1528	5696	5186	5184	8256	6716
3	4096	61440	45060	44924	44856	136704	128514	128480	198144	173404

Table 2-4 presents the number of real multiplications and additions required for the FFT of complex data and, therefore, gives a good picture of how much arithmetic reduction can be achieved using the various ideas of this section. M1 and A1 are the number of real data multiplications and additions for the straight-forward one-butterfly program in Figure 2-12 and for the counterparts for radix 4, 8, and 16. M2 and A2 are the numbers of operations for a two-butterfly program such as in Figure 2-13 for various radices. M3 and A3 correspond to a three-butterfly FFT which removes all the multiplies by one and j , and reduces some of the multiplications of the form in (2.101). This is one of the most efficient forms for a general-purpose machine. M5 and A3 correspond to a five-butterfly FFT which removes all unnecessary multiplications except those that could be removed by use of the three real multiplications per complex multiplication algorithm of (2.102) and (2.103). Notice that while the number of multiplications always decreases as the radix increases, the number of additions reaches a minimum and then increases. Also note that the savings by going from two to three butterflies is not as great as going from one to two, and that the savings decreases for the larger radices. The effects of special butterflies on quantization error is discussed in [18].

2.2.12 The Prime-Factor FFT Algorithm (PFA)

This section includes the algorithms that result from using the prime-factor index map (PFM) to calculate the DFT in an efficient way. The PFM was applied independently by Good and by Thomas [6,8] but did not prove useful until further developed by Winograd [8,9,11]. There are two forms of algorithms that use the PFM: the straight-forward use of Winograd's short prime-length DFTs to the decomposition of (2.81) which is called the prime-factor algorithm (PFA) [12,13], and the "nested" version which is called the Winograd Fourier Transform Algorithm (WFTA) and is covered in the next Section 2.2.13 [8,9,13].

The theory of the PFA is developed here for a two-factor example with $N = N1 * N2$. The particular time and frequency index maps used are (2.64) and (2.69), which are

$$n = \langle K_1 n_1 + K_2 n_2 \rangle_N \qquad (2.106)$$

$$k = \langle K_3 k_1 + K_4 k_2 \rangle_N$$

with the conditions for a unique mapping from (2.67) being

$$K_1 = aN_2 \text{ and } K_2 = bN_1 \qquad (2.107)$$

$$\text{and} \quad K_3 = cN_2 \text{ and } K_4 = dN_1$$

The conditions for uncoupling of row and column calculations are given by

$$\langle K_1 K_4 \rangle_N = 0 \text{ and } \langle K_2 K_3 \rangle_N = 0 \qquad (2.108)$$

and the requirements that cause the short sums to be length N1 and N2 DFTs are

$$\langle K_1K_3\rangle_N = N_2 \text{ and } \langle K_2K_4\rangle_N = N_1 \tag{2.109}$$

One set of coefficients that satisfy all these conditions is

$$a = b = 1 \tag{2.110}$$

and

$$K_3 = N_2\langle N_2^{-1}\rangle_{N_1} \text{ and } K_4 = N_1\langle N_1^{-1}\rangle_{N_2} \tag{2.111}$$

This results in simple coefficients for the time map but somewhat complicated ones for the frequency map. The frequency map is called the Chinese Remainder Theorem (CRT), and the result of the selection of coefficients in (2.111) makes

$$k_1 = \langle k\rangle_{N_1} \text{ and } k_2 = \langle k\rangle_{N_2} \tag{2.112}$$

Other maps are possible, but (2.110) makes K_1 and K_2 as small as possible which simplifies programming the indexing. The two maps of (2.110) and (2.111) and the CRT are discussed in [8,9,11,13]. Both maps are written as

$$\begin{aligned} n &= \langle N_2n_1 + N_1n_2\rangle_N \\ k &= \langle K_3k_1 + K_4k_2\rangle_N \end{aligned} \tag{2.113}$$

In contrast to the case in (2.86) using the CFM, the indicated residue reduction must explicitly be carried out. The DFT becomes

$$\hat{X}(k_1,k_2) = \sum_{n_2=0}^{N_2-1} \sum_{n_1=0}^{N_1-1} \hat{x}(n_1,n_2)\, W_{N_1}^{n_1k_1}\, W_{N_2}^{n_2k_2} \tag{2.114}$$

This is a pure two-dimensional DFT of the same form as the example in (2.79) and (2.81). Again, in contrast to the Cooley-Tukey FFT which uses the CFM in Section 2.2.9, there is no TF term, and the order of the summation can be interchanged.

This is illustrated in Figure 2-15 where each of the short length-N_i DFTs are calculated by a Winograd-type prime-length algorithm. Each of these short algorithms consists of a set of additions, followed by a set of multiplications, and then followed by another set of additions.

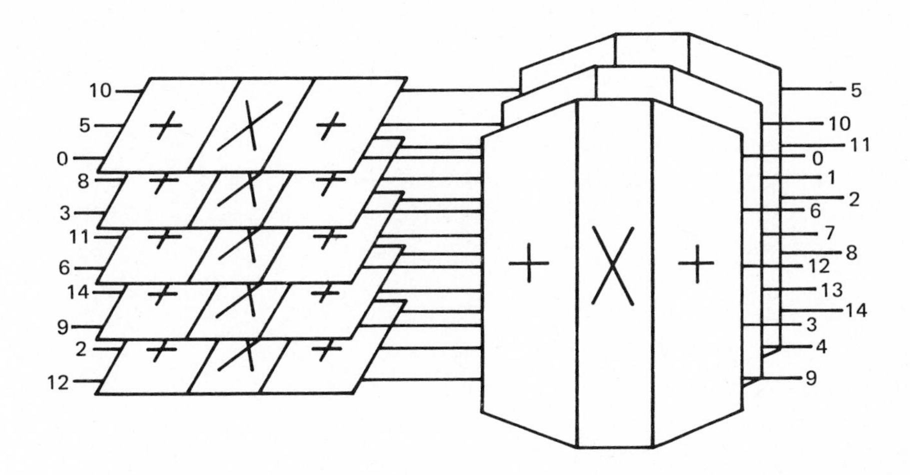

FIGURE 2-15. A PRIME-FACTOR FFT WITH TWO FACTORS

If the length N has three relatively prime factors given by

$$N = N_1 N_2 N_3 \tag{2.115}$$

the time index map becomes

$$n = \langle N_2 N_3 n_1 + N_1 N_3 n_2 + N_1 N_2 n_3 \rangle_N \tag{2.116}$$

and the DFT becomes

$$\hat{X} = \sum_{n_3=0}^{N_3-1} \sum_{n_2=0}^{N_2-1} \sum_{n_1=0}^{N_1-1} \hat{x}\, W_{N_1}^{n_1 k_1} W_{N_2}^{n_2 k_2} W_{N_3}^{n_3 k_3} \tag{2.117}$$

Four factors would give four nested summations. These sums can be performed in any order.

The programming of the PFM indexing is more efficient if carried out in a slightly different way from the CFM indexing. Rather than calculating a three-dimensional DFT as in (2.117), a sequence of two-dimensional calculations is performed. First, the N_1 part of an N_1 by N_2N_3 DFT is calculated, followed by the N_2 part of a N_2 by N_1N_3 DFT, and finally the N_3 part of a N_3 by N_1N_2 DFT is done.

For the first calculation, (2.117) looks like

$$\hat{X} = \sum_{n_4=0}^{N_3N_2-1} \sum_{n_1=0}^{N_1-1} \hat{x}\, W_{N_1}^{n_1k_1}\, W_{N_2N_3}^{n_4k_4} \qquad (2.118)$$

where n_4 and k_4 go from 0 to $(N_2N_3 - 1)$. Note that the inner sum in (2.118) is exactly the same as the inner sum in (2.117). After this inner sum is carried out, the problem is reformulated as

$$\hat{X} = \sum_{n_4=0}^{N_1N_3-1} \sum_{n_2=0}^{N_2-1} \hat{x}\, W_{N_2}^{n_2k_2}\, W_{N_1N_3}^{n_4k_4} \qquad (2.119)$$

where for this sum, n_4 and k_4 go from 0 to $(N_1N_3 - 1)$, and the sum over n_2 is exactly the same as in (2.117). The final sum is of the same sort and is also the same as in (2.117). This calculation gives the same answer for the DFT as would have been obtained from (2.117) but the indexing is simpler to program.

In the case of the Cooley-Tukey FFT, each stage is a DFT with a length equal to the radix and is called a butterfly. Because each stage has the same butterfly, only one is programmed, and it is used repeatedly in nested loops. For the prime-factor DFT, each stage has a different length DFT or butterfly; therefore, an M-stage algorithm must have M different butterflies, now called "modules". Figure 2-16 contains the FORTRAN statements for the indexing part of a prime-factor FFT and Chapter 4 has a complete PFA program.

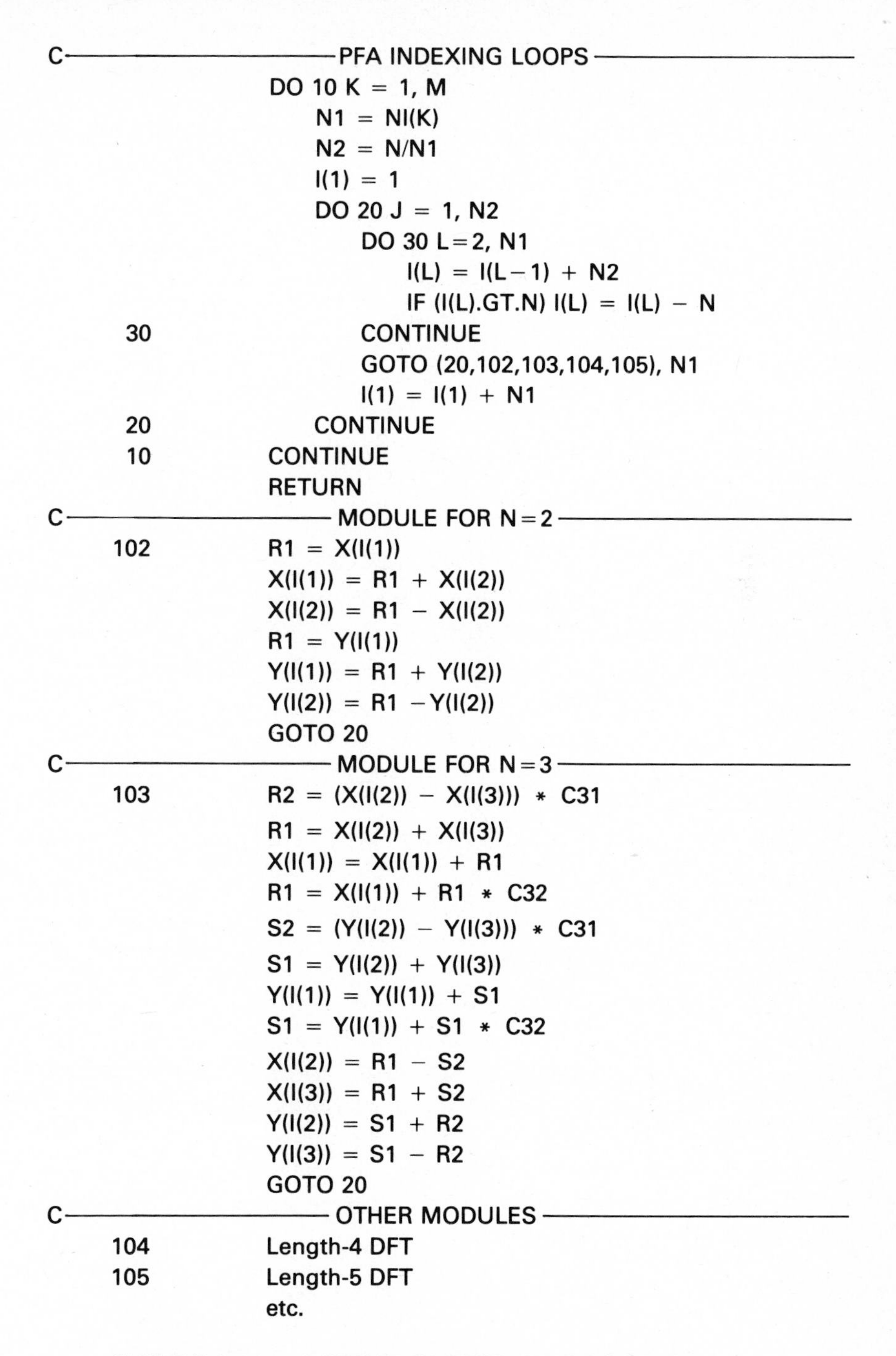

```
C-------------------PFA INDEXING LOOPS-------------------
            DO 10 K = 1, M
                N1 = NI(K)
                N2 = N/N1
                I(1) = 1
                DO 20 J = 1, N2
                    DO 30 L=2, N1
                        I(L) = I(L-1) + N2
                        IF (I(L).GT.N) I(L) = I(L) - N
   30               CONTINUE
                    GOTO (20,102,103,104,105), N1
                    I(1) = I(1) + N1
   20           CONTINUE
   10       CONTINUE
            RETURN
C-------------------MODULE FOR N=2-------------------
  102       R1 = X(I(1))
            X(I(1)) = R1 + X(I(2))
            X(I(2)) = R1 - X(I(2))
            R1 = Y(I(1))
            Y(I(1)) = R1 + Y(I(2))
            Y(I(2)) = R1 -Y(I(2))
            GOTO 20
C-------------------MODULE FOR N=3-------------------
  103       R2 = (X(I(2)) - X(I(3))) * C31
            R1 = X(I(2)) + X(I(3))
            X(I(1)) = X(I(1)) + R1
            R1 = X(I(1)) + R1 * C32
            S2 = (Y(I(2)) - Y(I(3))) * C31
            S1 = Y(I(2)) + Y(I(3))
            Y(I(1)) = Y(I(1)) + S1
            S1 = Y(I(1)) + S1 * C32
            X(I(2)) = R1 - S2
            X(I(3)) = R1 + S2
            Y(I(2)) = S1 + R2
            Y(I(3)) = S1 - R2
            GOTO 20
C-------------------OTHER MODULES-------------------
  104       Length-4 DFT
  105       Length-5 DFT
            etc.
```

FIGURE 2-16. A PRIME-FACTOR FFT PROGRAM

This program takes input data with the real part in the array X(I) and the imaginary part in array Y(I) and calculates a DFT in place. The length N has M relatively prime factors which are in the array NI(K), such that N = NI(1) * NI(2) * ... * NI(M).

In the FORTRAN program, the outer DO 10 loop steps through the M stages of the algorithm, sequentially forming a two-dimensional problem as was shown in (2.118) and (2.119), with N1 taking on the various values and N2 being N/N1. The DO 20 loop calculates the N2 length-N1 DFTs. The locations of the proper input data points are calculated in the DO 30 loop which implements the input index map of

$$n = \langle N_2 n_1 + N_1 n_2 \rangle_N \tag{2.120}$$

In Figure 2-16, the first statement in the DO 30 loop indexes n_1, the statement just preceding line 20 indexes n_2, and the IF statement carries out the residue reduction.

The computed GOTO statement jumps to the appropriate module where the length-N1 DFT is calculated and put in the locations of the data. The GOTO 20 statement at the end of the module jumps back into the loop to continue with the calculations. It is poor programming practice to jump out of and into loops, but it is done here to present an uncluttered picture of the indexing part of a prime-factor program.

After all M stages are complete, the length-N DFT is finished and is located in arrays X and Y; however, the order is scrambled because only the input index maps of (2.120) and (2.110) were used, as was discussed in Section 2.2.8.

There are five approaches to dealing with the scrambled output of the PFA. First, there are some applications where the output does not have to be unscrambled. The spectrum can be used in scrambled order. This is sometimes possible when the FFT or PFA is used for high-speed convolution, discussed in Chapter 3. Second, an unscrambler stage can be added after the main PFA just as the bit-reverse counter is added to the Cooley-Tukey FFT. A simple form of unscrambler is listed in Chapter 4 and is not in-place. Third, the unscrambling can be done while the DFT is being calculated in each module. This is probably the fastest implementation, but it must be written specifically for a given length and rewritten if the length is changed. That approach is discussed in [12]. A fourth similar method is to unscramble the output by properly choosing the multiplier constants in the modules [14]. The fifth method is to use separate pointers for the input and output in each module; this is discussed in [15] and is listed in Chapter 4.

One method for evaluating a numerical algorithm is to count the necessary arithmetic operations on the data. That is certainly not all that takes time, but it is a first approximation. As was done for the Cooley-Tukey, a count of multiplications and additions is made. The number of multiplications and additions for the short Winograd type modules is given in Table 2-5.

TABLE 2-5. NUMBER OF REAL MULTIPLICATIONS AND ADDITIONS FOR A LENGTH-N DFT OF REAL DATA (DOUBLE FOR COMPLEX DATA)

N	MULTIPLIES	ADDS
2	0	2
3	2	6
4	0	8
5	5	17
7	8	36
8	2	26
9	10	42
11	20	84
13	20	94
16	10	74
17	35	157
19	38	186
25	66	210

The code for the modules is given in Chapter 4. Note that both the number of multiplications and additions as a function of length is rather erratic. Lengths that are a power of two are the most efficient. Lengths of 11 ,13, 17, 19, and 25 are rather inefficient, particularly 11, 17, and 19. Longer modules can be derived but they become progressively less efficient, and the code becomes very long.

The number of real multiplications and additions for the PFA can easily be calculated from Table 2-5. In the first stage of a four-factor PFA, there are N4 * N3 * N2 length-N1 DFTs which have N4 * N3 * N2 times the number of adds and multiplies for the length-N1 DFT in Table 2-5. This is repeated for the four stages and the results for complex data are given in Table 2-6.

TABLE 2-6. REAL OPERATION COUNT FOR A FOUR-FACTOR COMPLEX PRIME-FACTOR FFT

N	=	N1 *	N2 *	N3 *	N4	MULTIPLIES	ADDS	M+A
1	=	1	1	1	1	0	0	0
2	=	2	1	1	1	0	4	4
3	=	1	3	1	1	4	12	16
4	=	4	1	1	1	0	16	16
5	=	1	1	1	5	10	34	44
6	=	2	3	1	1	8	36	44
7	=	1	1	7	1	16	72	88
8	=	8	1	1	1	4	52	56
9	=	1	9	1	1	20	84	104
10	=	2	1	1	5	20	88	108
12	=	4	3	1	1	16	96	112
14	=	2	1	7	1	32	172	204

TABLE 2-6. REAL OPERATION COUNT FOR A FOUR-FACTOR COMPLEX PRIME-FACTOR FFT (CONTINUED)

N	=	N1 *	N2 *	N3 *	N4 ,	MULTIPLIES	ADDS	M+A
15	=	1	3	1	5	50	162	212
16	=	16	1	1	1	20	148	168
18	=	2	9	1	1	40	204	244
20	=	4	1	1	5	40	216	256
21	=	1	3	7	1	76	300	376
24	=	8	3	1	1	44	252	296
28	=	4	1	7	1	64	400	464
30	=	2	3	1	5	100	384	484
35	=	1	1	7	5	150	598	748
36	=	4	9	1	1	80	480	560
40	=	8	1	1	5	100	532	632
42	=	2	3	7	1	152	684	836
45	=	1	9	1	5	190	726	916
48	=	16	3	1	1	124	636	760
56	=	8	1	7	1	156	940	1096
60	=	4	3	1	5	200	888	1088
63	=	1	9	7	1	284	1236	1520
70	=	2	1	7	5	300	1336	1636
72	=	8	9	1	1	196	1140	1336
80	=	16	1	1	5	260	1284	1544
84	=	4	3	7	1	304	1536	1840
90	=	2	9	1	5	380	1632	2012
105	=	1	3	7	5	590	2214	2804
112	=	16	1	7	1	396	2188	2584
120	=	8	3	1	5	460	2076	2536
126	=	2	9	7	1	568	2724	3292
140	=	4	1	7	5	600	2952	3552
144	=	16	9	1	1	500	2676	3176
168	=	8	3	7	1	692	3492	4184
180	=	4	9	1	5	760	3624	4384
210	=	2	3	7	5	1180	4848	6028
240	=	16	3	1	5	1100	4812	5912
252	=	4	9	7	1	1136	5952	7088
280	=	8	1	7	5	1340	6604	7944
315	=	1	9	7	5	2050	8322	10372
336	=	16	3	7	1	1636	7908	9544
360	=	8	9	1	5	1700	8148	9848
420	=	4	3	7	5	2360	10536	12896
504	=	8	9	7	1	2524	13164	15688
560	=	16	1	7	5	3100	14748	17848
630	=	2	9	7	5	4100	17904	22004
720	=	16	9	1	5	3940	18276	22216
840	=	8	3	7	5	5140	23172	28312
1008	=	16	9	7	1	5804	29100	34904
1260	=	4	9	7	5	8200	38328	46528
1680	=	16	3	7	5	11540	50964	62504
2520	=	8	9	7	5	17660	82956	100616
5040	=	16	9	7	5	39100	179772	218872

The efficiency of a total PFA program depends on the efficiency of the particular modules and the number of them. Note the number of operations listed in Table 2-6 for lengths 35 and 36, and for 140 and 144. This variation is more pronounced for the shorter total lengths, thus indicating that the choice of lengths and factors can be very important.

Although a particular order is indicated in Table 2-6, the number of operations is independent of the order of the stages. The M + A column is a sum of the number of multiplications and additions and is given as a rough measure of the total arithmetic required. It has real meaning only if the multiply and add times are independent and equal. That occurs for some general-purpose computers with hardware floating-point arithmetic. The M + A number is not meaningful for most microprocessors or signal processors, such as the TMS32010.

Tables 2-4 and 2-6 should be studied and compared to understand the two different approaches and choose the most appropriate one.It should be remembered that simple multiply and add counts are not enough. Indexing and data transfers must be accounted for, and some algorithms fit some hardware better than others.

This section has discussed the indexing and organization of the PFA in some detail but has glossed over the details of the short modules themselves. The theory for deriving the short modules is somewhat complicated and difficult to concisely present. These details can be found in [8,9,10,16]. Fortunately, it is not necessary to understand the details of the modules to effectively use them.

2.2.13 The Winograd Fourier Transform Algorithm (WFTA)

In 1976, S. Winograd introduced a new approach to the efficient calculation of the DFT that uses substantially fewer multiplications than the traditional Cooley-Tukey FFT, but at the expense of more additions [4,8,9,13]. This algorithm uses the PFM for indexing (discussed in Section 2.2.7), converts the resulting short DFTs into cyclic convolution by Rader's technique (see Section 2.2.5), calculates the short convolution by a new optimal method based on a polynomial Chinese remainder theorem (CRT) [8,9], and reorders the operations so that all the multiplications are "nested" together in the center of the algorithm to reduce their number. The WFTA is exactly the same as the PFA of the previous Section 2.2.11, but with the final nesting of the multiplications reducing the number performed at execution of the program.

In order to effectively use the WFTA, it is necessary to understand more of the theory behind these short DFT algorithms than was necessary for the PFA. The idea is to operate on the length-N input data vector by simple additions to form a set of M intermediate variables where, in general, $M > N$. These intermediate variables are multiplied by constants which are derived from a combination of the real and imaginary parts of W in the DFT. These M products are then combined with simple additions to give the N values of the DFT. The three multiplication algorithms in (2.102) and (2.103) have this form.

The derivation of this add-multiply-add structure comes from considering the input data as a polynomial with the data values as coefficients. This polynomial is decomposed by a polynomial residue reduction modulo the polynomial $(z^{N-1} - 1)$ which is similar to normal integer residue reduction modulo an integer. After some further reduction, these residues are multiplied by the residues obtained from reducing a polynomial formed from W. Finally, the product residues are recombined by a polynomial form of the Chinese remainder theorem to give the DFT. The reduction gives the first additions, and the CRT gives the final additions to the total structure. The necessary number theory and polynomial theory together with the application to the DFT problem are presented in [8,9]. A good list of algorithms for short DFT modules based on this theory is contained in [9], and a straight-forward procedure for deriving others, plus a listing of some of the longer modules, is in [10].

The same index map as was used for the PFA in (2.113) is used for the WFTA. The two-factor uncoupled DFT form (2.114) is

$$\hat{X}(k_1,k_2) = \sum_{n_2=0}^{N_2-1} \sum_{n_1=0}^{N_1-1} \hat{x}(n_1,n_2)\, W_{N_1}^{n_1k_1}\, W_{N_2}^{n_2k_2} \tag{2.121}$$

Recall that the inner sum is a set of identical length-N_1 DFTs along the N_2 rows of the $\hat{x}(n_1,n_2)$ array, and the outer sum is a set of length-N_2 DFTs along the N_1 columns of the intermediate array which is a function of (k_1,n_2). It was shown by Winograd that these operations commute, i.e., can be done in any order.

If the short DFT modules are factored into three sequencially applied operators, they will have an addition operator, followed by a multiplication operator, and finally followed by another addition operator. This is illustrated in Figure 2-17.

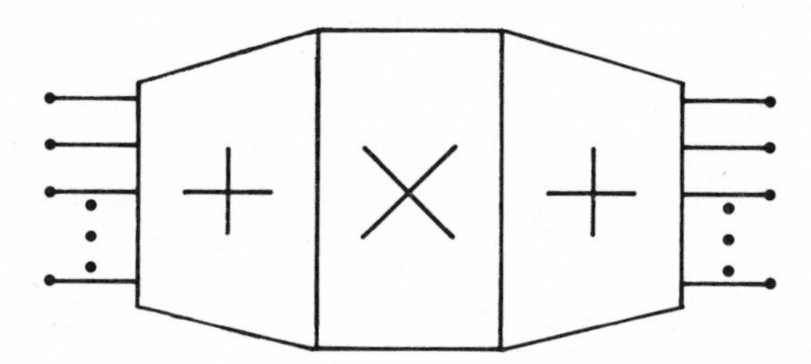

FIGURE 2-17. A SHORT WINOGRAD DFT MODULE

The taller center section in Figure 2-17 indicates that the number of multiplications is in general larger than the number of data points. An example flowgraph for a length-5 module is presented in Figure 2-18 to show the add-multiply-add structure.

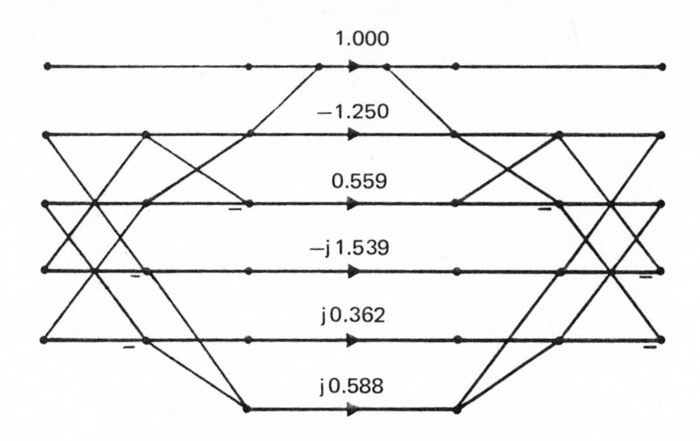

FIGURE 2-18. A LENGTH-5 WINOGRAD DFT MODULE FLOWGRAPH

This picture of the DFT module is combined with the picture of the uncoupling of the row and column calculations that occurs with the index mapping in Figures 2-8 and 2-9 to give Figure 2-15 in Section 2.2.12. Using the ability to commute or reorder the operations, the multiplication stages can all be made adjacent as shown in Figure 2-19.

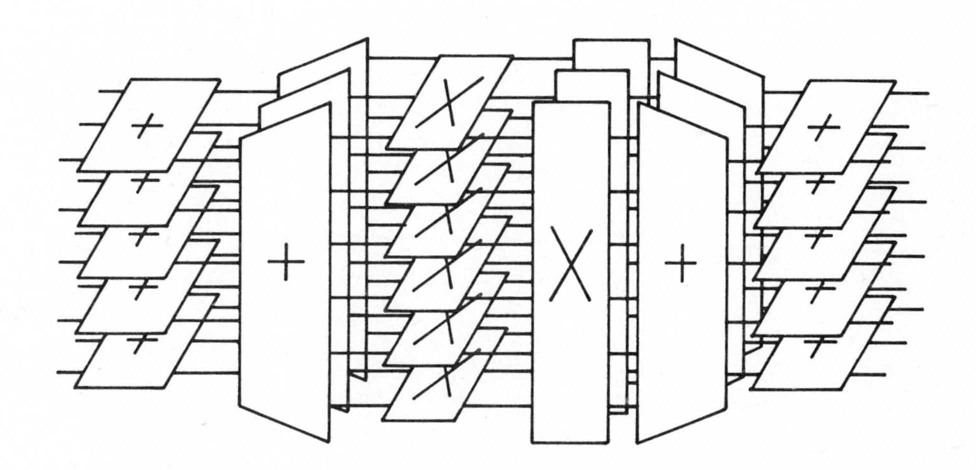

FIGURE 2-19. THE WFTA WITH NESTED MULTIPLICATIONS

In execution, a variable would not first be multiplied by one constant and then by another in the center two stages. Instead,the two multiplication stages would be pre-multiplied into one stage of constants by which the variables then are multiplied. This moving the multiplications together in the center of the algorithm is called "nesting" and is responsible for the small number of multiplications in the WFTA.

Note that the rectangular structure of the addition operators causes an "expansion" of the data in the center of the algorithm. Rather than the 15 data points, there are

18 intermediate values. This expansion is a major problem with implementing the nested WFTA because it prevents a straight-forward in-place operation and causes an increase in the number of additions and the number of multiplier constants that must be stored.

A simple length-12 two-factor program is given in Figure 2-20 that implements the PFA using multidimensional arrays rather than the approach in Figure 2-16. The program in Figure 2-20 is more complicated, but it is in a form that can be extended to WFTA where the form of Figure 2-16 cannot.

```
C-------------------- MULTIPLIER COEFFICIENTS ------------
      DATA C31, C32 / -.86602540, -1.50/
C-------------------- INPUT INDEX MAP --------------------
      L = 1
      DO 2 J = 1, N2
         DO 1 I = 1, N1
            XA(I,J) = X(L)
            YA(I,J) = Y(L)
            L = L + N2
            IF (L.GT.N) L = L - N
1        CONTINUE
         L = L + N1
         IF (L.GT.N) L = L - N
2     CONTINUE
C-------------------- LENGTH-3 ROW DFT'S ------------------
      DO 3 J = 1, N2
         R2 = (XA(2,J) - XA(3,J)) * C31
         R1 = XA(2,J) + XA(3,J)
         XA(1,J) = XA(1,J) + R1
         R1 = XA(1,J) + R1 * C32
C
         S2 = (YA(2,J) - YA(3,J)) * C31
         S1 = YA(2,J) + YA(3,J)
         YA(1,J) = YA(1,J) + S1
         S1 = YA(1,J) + S1 * C32
C
         XA(2,J) = R1 - S2
         XA(3,J) = R1 + S2
         YA(2,J) = S1 + R2
         YA(3,J) = S1 - R2
3     CONTINUE
```

```
C—————————— LENGTH-4 COLUMN DFT'S ——————————
          DO 4 I = 1, N1
             R1 = XA(I,1) + XA(I,3)
             R2 = XA(I,1) - XA(I,3)
             T = XA(I,2) + XA(I,4)
             R4 = XA(I,2) - XA(I,4)
             XA(I,1) = R1 + T
             XA(I,3) = R1 - T
             R1 = YA(I,1) + YA(I,3)
             S2 = YA(I,1) - YA(I,3)
             T = YA(I,2) + YA(I,4)
             S4 = YA(I,2) - YA(I,4)
C
             XA(I,2) = R2 + S4
             XA(I,4) = R2 - S4
             YA(I,1) = R1 + T
             YA(I,3) = R1 - T
             YA(I,2) = S2 - R4
             YA(I,4) = S2 + R4
    4     CONTINUE
C—————————— OUTPUT INDEX MAP ——————————
          L = 1
          DO 6 J = 1, N2
             DO 5 I = 1, N1
                X(L) = XA(I,J)
                Y(L) = YA(I,J)
                L = L + K3
                IF (L.GT.N) L = L - N
    5        CONTINUE
             L = L + K4
             IF (L.GT.N) L = L - N
    6     CONTINUE
```

FIGURE 2-20. THE PFA WITH MULTIDIMENSIONAL ARRAYS

This program takes inputs with real parts in X and imaginary parts in Y and, using the index map of (2.106) and 2.113), creates the two-dimensional real and imaginary part arrays, XA and YA. Then the four length-3 DFTs are calculated followed by the three length-4 DFTs. The final step implements the output map (2.106) and (2.111) by placing the DFT in proper order back into the orginal X and Y locations.

This program is not in-place since the XA and YA arrays are required. The structure can be extended to more factors and has the basic form of the program by McClellan [4,8]. In Figure 2-21, the multiplications are nested together in the center of the algorithm, as shown in Figure 2-19.

```
C---------------------- MULTIPLIER COEFFICIENTS ----------------------
              DATA C31, C32 / -.86602540, -1.50/
C---------------------- INITIALIZE—CALCULATE MULTIPLIER ARRAY —
              DO 30 J = 1, N2
                 M(1,J) = 1
                 M(2,J) = C32
                 M(3,J) = C31
   30         CONTINUE
C---------------------- INPUT INDEX MAP ----------------------
              L = 1
              DO 2 J= 1, N2
                 DO 1 I = 1, N1
                    XA(I,J) = X(L)
                    YA(I,J) = Y(L)
                    L = L + N2
                    IF (L.GT.N) L = L - N
    1            CONTINUE
                 L = L + N1
                 IF (L.GT.N) L = L - N
    2         CONTINUE
C---------------------- LENGTH-3 PREWEAVE ----------------------
              DO 3 J = 1, N2
                 T = XA(2,J) + XA(3,J)
                 XA(3,J) = XA(2,J) - XA(3,J)
                 XA(1,J) = XA(1,J) + T
                 XA(2,J) = T
C
                 T = YA(2,J) + YA(3,J)
                 YA(3,J) = YA(2,J) - YA(3,J)
                 YA(1,J) = YA(1,J) + T
                 YA(2,J) = T
    3         CONTINUE
C---------------------- LENGTH-4 PREWEAVE ----------------------
              DO 4 I = 1, N1
                 T = XA(I,1) - XA(I,3)
                 XA(I,1) = XA(I,1) + XA(I,3)
                 XA(I,3) = T
                 T = XA(I,2) - XA(I,4)
                 XA(I,2) = XA(I,2) + XA(I,4)
                 XA(I,4) = T
C
                 T = YA(I,1) - YA(I,3)
                 YA(I,1) = YA(I,1) + YA(I,3)
                 YA(I,3) = T
                 T = YA(I,2) - YA(I,4)
                 YA(I,2) = YA(I,2) + YA(I,4)
                 YA(I,4) = T
    4         CONTINUE
```

```
C------------------------ NESTED MULTIPLIES ------------------------
          DO 6 J = 1, N2
             DO 5 I = 1, N1
                XA(I,J) = XA(I,J) * M(I,J)
                YA(I,J) = YA(I,J) * M(I,J)
    5        CONTINUE
    6     CONTINUE
C------------------------ LENGTH-3 POSTWEAVE ------------------------
          DO 7 J = 1, N2
             XA(2,J) = XA(2,J) + XA(1,J)
             YA(2,J) = YA(2,J) + YA(1,J)
             XT = XA(2,J) + YA(3,J)
             YT = YA(2,J) - XA(3,J)
             XA(2,J) = XA(2,J) - YA(3,J)
             YA(2,J) = YA(2,J) + XA(3,J)
             XA(3,J) = XT
             YA(3,J) = YT
    7     CONTINUE
C------------------------ LENGTH-4 POSTWEAVE ------------------------
          DO 8 I = 1, N2
             XT = XA(I,1) - XA(I,2)
             XA(I,1) = XA(I,1) + XA(I,2)
             YT = YA(I,1) - YA(I,2)
             YA(I,1) = YA(I,1) + YA(I,2)
C
             XA(I,2) = XA(I,3) + YA(I,4)
             XXT = XA(I,3) - YA(I,4)
             XA(I,3) = XT
             YA(I,2) = YA(I,3) - XA(I,4)
             YA(I,4) = YA(I,3) + XA(I,4)
             XA(I,4) = XXT
             YA(I,3) = YT
    8     CONTINUE
C------------------------ OUTPUT INDEX MAP ------------------------
          L = 1
          DO 10 J = 1, N2
             DO 9 I = 1, N1
                X(L) = XA(I,J)
                Y(L) = YA(I,J)
                L = L + K3
                IF (L.GT.N) L = L - N
    9        CONTINUE
             L = L + K4
             IF (L.GT.N) L = L - N
   10     CONTINUE
```

FIGURE 2-21. A NESTED WFTA FOR N = 3 * 4 = 12

A comparison of Figures 2-20 and 2-21 shows how the length-3 and length-4 DFT modules were split into a preweave addition section, followed by a multiplication section, and then followed by a postweave addition section (see Figures 2-18 and 2-19).

The appropriate multiplication constants are stored in the array MA. In general, this array will have a size of M(N1) * M(N2) * M(N3) * ..., where M(N) is the number of multiplications of a length-N DFT module and indicates the amount of expansion of a module.

The number of additions is somewhat difficult to count because it depends on the order of the first addition stages (called preweave stages because of the appearance of their flowgraphs) and the order of the last addition stages (postweave stages). The additions for the two-factor example are given by

$$\text{additions} = N_2A_1 + M_1A_2 \tag{2.122}$$

where A_i is the number of additions in the i-th stage, and M_i is the number of multiplications in the i-th stage. If the order of the preweave stages were reversed, the number would be

$$\text{additions} = N_1A_2 + M_2A_1 \tag{2.123}$$

If the number of total additions is to be minimized, a proper ordering must be found for each length N. The multiplication and add numbers for each of the short Winograd modules is given in Table 2-7. Note these numbers are different from those in Table 2-5 for the PFA. Since for the WFTA all the multiplications are nested together, even multiplications by one in the short modules must be counted. For the PFA, that is not the case.

TABLE 2-7. NUMBER OF REAL MULTIPLICATIONS AND ADDITIONS FOR A LENGTH-N DFT OF REAL DATA USED WITH THE WFTA (DOUBLE THE NUMBERS FOR COMPLEX DATA)

N	TOTAL MULTIPLIES	MULTIPLIES BY ONE	ADDS
2	2	2	2
3	3	1	6
4	4	4	8
5	6	1	17
7	9	1	36
8	8	6	26
9	11	1	43
16	18	8	74

These numbers are consistant with the WFTA program written by McClellan and published in [4,8]. The other lengths given in Table 2-5 have not been implemented on the WFTA although they easily could be if desired.

The total number of multiplications and additions for the WFTA are given in Table 2-8. The ordering is optimal for almost all of the lengths and near optimal for the remainder. This is the ordering used in [4,8]. The total number of multiplications is given by the product of the multiplications in each stage. Some reduction can be achieved by removing multiplications by one but that reduction is difficult to program. The reduced number of multiplications is also given in Table 2-8.

TABLE 2-8. REAL OPERATION COUNT FOR A FOUR-FACTOR WFTA WITH COMPLEX DATA AND USING McCLELLAN'S ORDERING

N	=	N1 *	N2 *	N3 *	N4	MULTIPLICATIONS TOTAL	REDUCED	ADDS	M+A
1	=	1	1	1	1	2	0	0	2
2	=	2	1	1	1	4	0	4	8
3	=	1	3	1	1	6	4	12	18
4	=	4	1	1	1	8	0	16	24
5	=	1	1	1	5	12	10	34	46
6	=	2	3	1	1	12	8	36	48
7	=	1	1	7	1	18	16	72	90
8	=	8	1	1	1	16	4	52	68
9	=	1	9	1	1	22	20	86	108
10	=	2	1	1	5	24	20	88	112
12	=	4	3	1	1	24	16	96	120
14	=	2	1	7	1	36	32	172	208
15	=	1	3	1	5	36	34	162	198
16	=	16	1	1	1	36	20	148	184
18	=	2	9	1	1	44	40	208	252
20	=	4	1	1	5	48	40	216	264
21	=	1	3	7	1	54	52	300	354
24	=	8	3	1	1	48	36	252	300
28	=	4	1	7	1	72	64	400	472
30	=	2	3	1	5	72	68	384	456
35	=	1	1	7	5	108	106	666	774
36	=	4	9	1	1	88	80	488	576
40	=	8	1	1	5	96	84	532	628
42	=	2	3	7	1	108	104	684	792
45	=	1	9	1	5	132	130	804	936
48	=	16	3	1	1	108	92	660	768
56	=	8	1	7	1	144	132	940	1084
60	=	4	3	1	5	144	136	888	1032
63	=	1	9	7	1	198	196	1394	1592
70	=	2	1	7	5	216	212	1472	1688

TABLE 2-8. REAL OPERATION COUNT FOR A FOUR-FACTOR WFTA WITH COMPLEX DATA AND USING McCLELLAN'S ORDERING (CONTINUED)

N	=	N1 * N2 * N3 * N4				MULTIPLICATIONS TOTAL	REDUCED	ADDS	M+A
72	=	8	9	1	1	176	164	1156	1332
80	=	16	1	1	5	216	200	1352	1568
84	=	4	3	7	1	216	208	1536	1752
90	=	2	9	1	5	264	260	1788	2052
105	=	1	3	7	5	324	322	2418	2742
112	=	16	1	7	1	324	308	2332	2656
120	=	8	3	1	5	288	276	2076	2364
126	=	2	9	7	1	396	392	3040	3436
140	=	4	1	7	5	432	424	3224	3656
144	=	16	9	1	1	396	380	2880	3276
168	=	8	3	7	1	432	420	3492	3924
180	=	4	9	1	5	528	520	3936	4464
210	=	2	3	7	5	648	644	5256	5904
240	=	16	3	1	5	648	632	5136	5784
252	=	4	9	7	1	792	784	6584	7376
280	=	8	1	7	5	864	852	7148	8012
315	=	1	9	7	5	1188	1186	10336	11524
336	=	16	3	7	1	972	956	8508	9480
360	=	8	9	1	5	1056	1044	8772	9828
420	=	4	3	7	5	1296	1288	11352	12648
504	=	8	9	7	1	1584	1572	14428	16012
560	=	16	1	7	5	1944	1928	17168	19112
630	=	2	9	7	5	2376	2372	21932	24308
720	=	16	9	1	5	2376	2360	21132	23508
840	=	8	3	7	5	2592	2580	24804	27396
1008	=	16	9	7	1	3564	3548	34416	37980
1260	=	4	9	7	5	4752	4744	46384	51136
1680	=	16	3	7	5	5832	5816	59064	64896
2520	=	8	9	7	5	9504	9492	99068	108572
5040	=	16	9	7	5	21384	21368	232668	254052

The WFTA allows the same lengths as the PFA, but the number of multiplications is smaller and the number of additions larger. The efficiency of the PFA depends on the efficiency of the individual modules. It is emphasized again that simple operation counts are not the total picture in evaluating an algorithm for possible application.

There are several improvements that could be made over this simple example. The program by McClellan should be consulted for a general four-factor WFTA. It is possible to merge the first stage that generates the multidimensional array and the first preweave into a single stage. It is also possible to merge the last preweave, the multiplication stage, and the first postweave stage, as well as the last postweave and the final array-to-output conversion. These changes will not reduce the arithmetic but will reduce the data transfer and indexing.

The WFTA has several disadvantages when compared to the Cooley-Tukey FFT and the PFA. A comparison of Figures 2-21, 2-15, and 2-11 shows the WFTA to be longer and more complicated. It requires more passes through the data (DO loops) and more index calculations. The data expansion prevents in-place calculation, and the nesting of the multiplications causes the number of stored multiplier coefficients to be much larger than for the PFA.

2.2.14 The DFT of Real Data

In many practical applications, the data to be processed are real, and the use of a general complex-data algorithm on real data is inefficient. This section shows how to convert a general algorithm for efficient processing of real data.

If the DFT of a length-2N real sequence, x(n), is desired, two length-N sequences are formed from the even and odd index samples.

$$\begin{aligned} u(n) &= x(2n) \\ v(n) &= x(2n+1) \end{aligned} \tag{2.124}$$

The two length-N DFTs can be calculated at the same time by one length-N complex DFT by forming an artificial complex signal from the two real signals, u(n) and v(n), as

$$h(n) = u(n) + jv(n)$$

The DFTs of u(n) and v(n) can be found from the DFT of h(n) using the properties of Table 2-1 which states that the DFT of u(n) has an even real part and an odd imaginary part and the DFT of jv(n) has an odd real part and an even imaginary part.

If The DFT of the original signal, x(n), can be found from the DFTs of u(n) and v(n) by using the sampling property from (2.23) which states

$$U(k) = \text{DFT}\{u(n)\} = [X(k) + X(k + N/2)]/2$$

The shift property from (2.18) gives

$$V(k) = DFT\{v(n)\} = [X(k)W^k + X(k + N/2)W^{k+N/2}]/2$$

Solving these two equations for X(k) gives

$$X(k) = U(k) + V(k)W^{-k} \tag{2.125}$$

The properties in Table 2-1 show that if x(n) is real, the values of X(k) for k = N,..., 2N − 1 are also given by (2.125); therefore, the length-2N DFT of x(n) can be found from two length-N DFTs if combined by (2.125).

$$H(k) = DFT\{h(n)\} = R(k) + jI(K)$$

then

$$U(k) = \text{even part of } R + j \text{ odd part of } I$$

$$V(k) = \text{even part of } I + j \text{ odd part of } R$$

These, together with (2.125), give the DFT of the real signal, x(n), from one complex half-length DFT and a small amount of extra calculation. This result and similar ones for other special cases are also covered in [4].

2.3 REFERENCES

1. Oppenheim, A.V. and Schafer, R.W., DIGITAL SIGNAL PROCESSING. Englewood Cliffs, NJ: Prentice-Hall, Inc., 1975.

2. Rabiner, L.R. and Gold, B., THEORY AND APPLICATION OF DIGITAL SIGNAL PROCESSING. Englewood Cliffs, NJ: Prentice-Hall, Inc., 1975.

3. Brigham, E.O., THE FAST FOURIER TRANSFORM. Englewood Cliffs, NJ: Prentice-Hall, Inc., 1974.

4. PROGRAMS FOR DIGITAL SIGNAL PROCESSING. New York, NY: IEEE Press, 1979.

5. Morris, L.R., DIGITAL SIGNAL PROCESSING SOFTWARE. Toronto, Canada: DSPSW, Inc., 1982, 1983.

6. Rabiner, L.R. and Rader, C.M. (editors), DIGITAL SIGNAL PROCESSING, Selected Reprints. New York, NY: IEEE Press, 1972.

7. Knuth, D.E., "SEMINUMERICAL ALGORITHMS," THE ART OF COMPUTER PROGRAMMING, Vol. 2. Reading, MA: Addison-Wesley, Inc., 1971, 1980.

8. McClellan, J.H. and Rader, C.M., NUMBER THEORY IN DIGITAL SIGNAL PROCESSING. Englewood Cliffs, NJ: Prentice-Hall, Inc., 1979.

9. Nussbaumer, H.J., FAST FOURIER TRANSFORM AND CONVOLUTION ALGORITHMS. Heidelberg, Germany: Springer Verlag, 1981, 1982.

10. Johnson, H.W. and Burrus, C.S., "Large DFT Modules: N = 11, 13, 17, 19, and 25," RICE UNIVERSITY EE DEPT. REPORT, No 8105, Houston, TX, 1982.

11. Burrus, C.S., "Index Mappings for Multidimensional Formulation of the DFT and Convolution," IEEE TRANS. ON ASSP, Vol. 25, June 1977, 239-242.

12. Burrus, C.S. and Eschenbacher, P.W., "An In-Place, In-Order Prime Factor FFT Algorithm," IEEE TRANS. ON ASSP, Vol. 29, August 1981, 806-817.

13. Kolba, D.P. and Parks, T.W. "A Prime Factor FFT Algorithm Using High Speed Convolution," IEEE TRANS. ON ASSP, Vol. 25, August 1977, 281-294.

14. Johnson, H.W. and Burrus, C.S., "Structures of DFT Algorithms," IEEE TRANS. ON ASSP, Vol. 33, February 1985.

15. Rothweiler, J.H., "Implementation of the In-Order Prime Factor FFT Algorithm," IEEE TRANS. ON ASSP, Vol. 30, February 1982, 105-107.

16. Elliot, D.F. and Rao, K.R., FAST TRANSFORMS: ALGORITHMS, ANALYSIS AND APPLICATIONS. New York, NY: Academic Press, 1982.

17. Blahut, R.E., FAST ALGORITHMS FOR DIGITAL SIGNAL PROCESSING. Reading, MA: Addison-Wesley, Inc., 1984.

18. Heute, U., "Results of a Deterministic Analysis of FFT Coefficient Errors," SIGNAL PROCESSING, North-Holland Publishing Co., Vol. 3, October 1981, 321-331.

19. Duhamel, P. and Hollmann, H., "Split Radix FFT Algorithm," ELECTRONIC LETTERS, Vol. 20, No. 1, January 5, 1984, 14-16.

20. Yavne, R., "An Economical Method for Calculating the Discrete Fourier Transform," PROCEEDINGS OF THE FALL JOINT COMPUTER CONFERENCE, 1968, 115-125.

Chapter 3

DISCRETE-TIME CONVOLUTION

Convolution is one of the most frequently used signal processing operations. The implementation of a FIR filter is usually done by direct convolution of the input signal with the impulse response of the filter. In spectral estimation, the basic operation of autocorrelation is simply the convolution of the signal with a reversed-time version of itself.

There is a close tie between discrete-time convolution and the DFT. As described in Chapter 2, the multiplication of two DFTs corresponds to the circular convolution of the corresponding time signals. Furthermore, as discussed in Section 2.2.5 under the description of "Rader's Convolution," it is possible to compute the DFT itself by converting it to a circular convolution.

This chapter begins with a description of the properties of convolution, with special emphasis on matrix descriptions of both linear and circular convolution. The calculation of a long convolution with a combination of short convolutions is discussed in some detail. Several methods for efficiently computing convolutions are described using matrix notation and FORTRAN statements.

3.1 PROPERTIES OF CONVOLUTION

The distinction between linear convolution and circular convolution is an important aspect of discrete-time convolution. This distinction is presented in terms of matrices and summations. The convolution properties necessary for segmentation of a long convolution into short convolutions are presented in this section. The connections between convolution and Fourier transforms are also described.

3.1.1 Linear Convolution

The linear convolution of two discrete-time signals, x(n) and h(n), is defined by the following equation:

$$y(n) = \sum_{n=-\infty}^{\infty} h(n-m)\, x(m) \tag{3.1}$$

The convolution of signals h and x in (3.1), which yields the signal y, is often written, as described in Section 2.1.3, using the symbol " * " to represent convolution as

$$y = h * x \tag{3.2}$$

The signals in (3.1) must be defined for all values of n, from $-\infty$ to ∞. If the signal x is equal to zero for n outside of the interval [N1,N2], it is called a "length-(N2 − N1 + 1) signal". For convenience in notation, this length will be denoted by N, i.e., N2 − N1 + 1 = N. Similarly, if the signal h is equal to zero for n outside of the interval [M1,M2], it is called a "length-(M2 − M1 + 1) signal". Again, to simplify notation, this length is denoted by M, i.e., M2 − M1 + 1 = M. For these two finite length signals, the resulting convolution y(n) will be equal to zero for

$$n < M1 + N1 \tag{3.3}$$

and

$$n > M2 + N2$$

Thus, the signal y(n) will be a length-(M2 + N2 − N1 − M1 + 1) signal. The length of the output y of the convolution is equal to one less than the sum of the lengths of the two signals h and x which are being convolved.

The convolution defined in (3.1) is shown graphically in Figure 3-1. The signal x is drawn as a rectangle while the signal h is drawn as a trapezoid. In this graphical interpretation of convolution, h(n − m) is interpreted as a time-reversed shifted version of h. As the index n of the output signal y increases, h(n − m) shifts to the right on the m axis. The output y(n) is zero for large negative values of n when h(n − m) is too far to the left to overlap x(m), and for large values of n when h(n − m) has shifted to the right so far that there is again no overlap of x(m) and h(n − m). The first nonzero output sample occurs for n − M1 = N1 or for n = M1 + N1. The last nonzero output sample occurs for n − M2 = N2 or for n = M2 + N2.

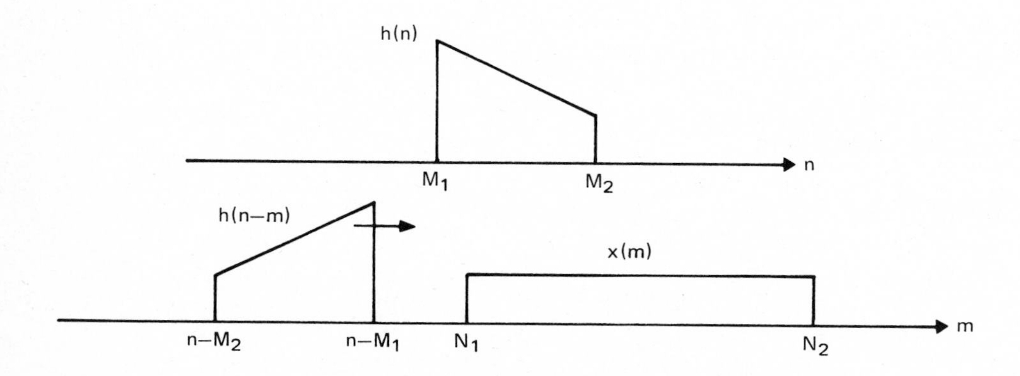

FIGURE 3-1. GRAPHICAL CONVOLUTION

3.1.1.1 A Matrix View of Convolution

The linear convolution, defined in (3.1) and illustrated in Figure 3-1, may be written as a matrix multiplication. The input signal x and the convolution output y are represented as column vectors, and x is multiplied with a convolution matrix H made up of columns which are shifted versions of the signal h. For $N1=0$, $N2=4$, $M1=0$, and $M2=2$, the matrix description of convolution is shown in (3.4).

$$\begin{bmatrix} y(0) \\ y(1) \\ y(2) \\ y(3) \\ y(4) \\ y(5) \\ y(6) \end{bmatrix} = \begin{bmatrix} h(0) & 0 & 0 & 0 & 0 \\ h(1) & h(0) & 0 & 0 & 0 \\ h(2) & h(1) & h(0) & 0 & 0 \\ 0 & h(2) & h(1) & h(0) & 0 \\ 0 & 0 & h(2) & h(1) & h(0) \\ 0 & 0 & 0 & h(2) & h(1) \\ 0 & 0 & 0 & 0 & h(2) \end{bmatrix} \begin{bmatrix} x(0) \\ x(1) \\ x(2) \\ x(3) \\ x(4) \end{bmatrix} \tag{3.4}$$

It is important to observe in (3.4) that the signal x has only five nonzero samples (x is a length-5 signal). The convolution sum (3.1), with the length-3 signal h in (3.4) and an *infinite-length* signal x, becomes

$$y(n) = \sum_{m=0}^{2} h(m)x(n-m) \qquad n=-\infty,\ldots,+\infty \tag{3.5}$$

The nth output y(n) is a weighted linear combination of the present and previous two samples for all values of n. Thus, in general, y(0) would depend on x(0), x(−1), and x(−2), and y(1) would depend on x(1), x(0), and x(−1). In (3.4), however, since x(−1) and x(−2) are zero,

$$y(0) = h(0) \cdot x(0) \tag{3.6}$$

and

$$y(1) = h(0) \cdot x(1) + h(1) \cdot x(0)$$

Similarly, since x(5) and x(6) are zero,

$$y(5) = h(0) \cdot 0 + h(1) \cdot x(4) + h(2) \cdot x(3) \tag{3.7}$$

and

$$y(6) = h(0) \cdot 0 + h(1) \cdot 0 + h(2) \cdot x(4)$$

3.1.1.2 *Transform Property*

The z transform of the linear convolution of two signals, x and h, is equal to the product of the z transform of x and the z transform of h. In other words, if

$$y = h * x \tag{3.8}$$

then

$$Y(z) = H(z)\, X(z) \tag{3.9}$$

It is important to note at this point that since (3.8) represents linear convolution, it is *not* true that the DFT of y is equal to the product of the DFTs of x and h. Since the DFT is a signal representation in terms of frequency samples (as described in Section 1.2.4), it corresponds to a periodic time signal. It is for this reason that the product of the DFTs corresponds to *cyclic* convolution (see Section 3.1.2), not linear convolution as defined by (3.1).

3.1.1.3 *Input Segmentation*

It is often necessary to work with short sections of a long input signal x when filtering. The impulse response of the filter h may typically have a length of 50 to 100, but the input signal may have a length in the thousands. There are two different procedures discussed in this book for segmenting the input, convolving short sections, and fitting everything back together for the correct output. The first method is called "output overlap-add" [1,2] in which the signal x (the input) is first written as a sum of length-B blocks.

$$x_m(n) = \begin{cases} x(n) & mB \le n \le mB + B - 1 \\ 0 & \text{else} \end{cases} \tag{3.10}$$

Substitution of (3.10) into (3.1) gives

$$y(n) = h * \sum_{m=-\infty}^{\infty} x_m(n) \tag{3.11}$$

With the definition of

$$y_m(n) = h * x_m(n) \tag{3.12}$$

equation (3.11) becomes

$$y(n) = \sum_{m=-\infty}^{\infty} y_m(n) \tag{3.13}$$

Each of the short convolutions in (3.12) involves a length-B input and a length-M signal h, giving a length-$(B+M-1)$ output signal y_m. These output blocks are overlapped and added, as indicated by (3.13) and shown in Figure 3-2.

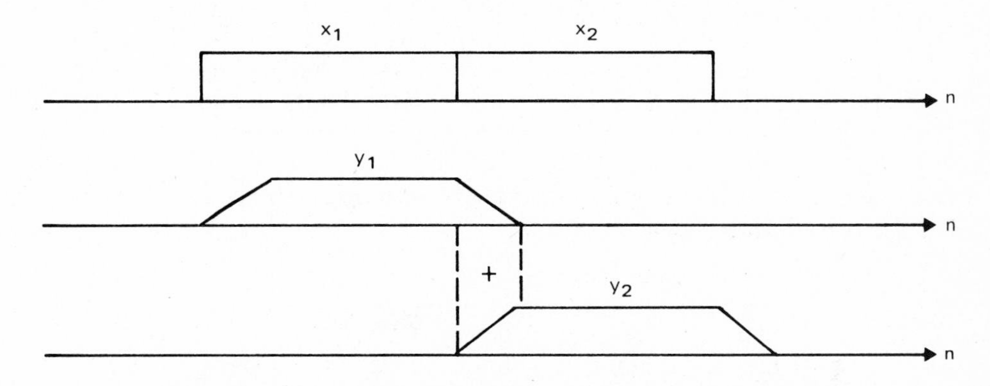

FIGURE 3-2. OUTPUT OVERLAP-ADD CONVOLUTION

The output overlap-add convolution may also be described using matrix notation as in (3.4). The length-8 output in (3.14) below may be calculated by sectioning the length-6 input into two sections of three samples each. The convolution of the length-3 signal h with each of these length-3 sections gives two length-5 outputs. These outputs are overlapped by two samples and added to give the output signal y of length-8.

$$
\begin{bmatrix} y(0) \\ y(1) \\ y(2) \\ y(3) \\ y(4) \\ y(5) \\ y(6) \\ y(7) \end{bmatrix}
=
\left[\begin{array}{ccc|ccc}
h(0) & 0 & 0 & 0 & 0 & 0 \\
h(1) & h(0) & 0 & 0 & 0 & 0 \\
h(2) & h(1) & h(0) & 0 & 0 & 0 \\
0 & h(2) & h(1) & h(0) & 0 & 0 \\
0 & 0 & h(2) & h(1) & h(0) & 0 \\
0 & 0 & 0 & h(2) & h(1) & h(0) \\
0 & 0 & 0 & 0 & h(2) & h(1) \\
0 & 0 & 0 & 0 & 0 & h(2)
\end{array}\right]
\begin{bmatrix} x(0) \\ x(1) \\ x(2) \\ x(3) \\ x(4) \\ x(5) \end{bmatrix}
\tag{3.14}
$$

The second method for sectioning the input is called "overlap-save" or "select-save". In this method, overlapping segments of the input signal are convolved with h, and the appropriate output values are selected and saved. The select-save algorithm takes advantage of the fact that the transient which results when convolving a length-M signal h(n) with a signal x(n) only lasts for $M-1$ samples. Thus only a length-$(M-1)$ overlap is required.The matrix form of convolution in (3.15) illustrates the idea. The two output sections, each of length-3, are computed using length-5 sections of the input x(n). The first two outputs are correct because the input x(n) is zero for $n=-1$ and $n=-2$. The portions of the convolution matrix H, which are used to calculate each of the two length-3 output sections, are enclosed in blocks in (3.15).

$$\begin{bmatrix} y(0) \\ y(1) \\ y(2) \\ y(3) \\ y(4) \\ y(5) \\ y(6) \\ y(7) \end{bmatrix} = \begin{bmatrix} h(0) & 0 & 0 & 0 & 0 & 0 & 0 & 0 \\ h(1) & h(0) & 0 & 0 & 0 & 0 & 0 & 0 \\ h(2) & h(1) & h(0) & 0 & 0 & 0 & 0 & 0 \\ 0 & h(2) & h(1) & h(0) & 0 & 0 & 0 & 0 \\ 0 & 0 & h(2) & h(1) & h(0) & 0 & 0 & 0 \\ 0 & 0 & 0 & h(2) & h(1) & h(0) & 0 & 0 \\ 0 & 0 & 0 & 0 & h(2) & h(1) & h(0) & 0 \\ 0 & 0 & 0 & 0 & 0 & h(2) & h(1) & h(0) \end{bmatrix} \begin{bmatrix} x(0) \\ x(1) \\ x(2) \\ x(3) \\ x(4) \\ x(5) \\ x(6) \\ x(7) \end{bmatrix} \quad (3.15)$$

The name "select-save" arises from the process of selecting only the last three of the five outputs of the convolution shown in (3.16).

$$\begin{bmatrix} \mathrm{tr} \\ \mathrm{tr} \\ y(5) \\ y(6) \\ y(7) \end{bmatrix} = \begin{bmatrix} h(0) & 0 & 0 & 0 & 0 \\ h(1) & h(0) & 0 & 0 & 0 \\ h(2) & h(1) & h(0) & 0 & 0 \\ 0 & h(2) & h(1) & h(0) & 0 \\ 0 & 0 & h(2) & h(1) & h(0) \end{bmatrix} \begin{bmatrix} x(3) \\ x(4) \\ x(5) \\ x(6) \\ x(7) \end{bmatrix} \quad (3.16)$$

In (3.16), a length-5 signal made from a section of the input is convolved with the length-3 signal h(n). The first two outputs, indicated by "tr" in (3.16) are transient, or start-up outputs. The transient only lasts $M-1 = 2$ samples in this example, so the remaining three outputs are the same as those that would be obtained when convolving a long signal x(n). Both the overlap-add and the select-save approaches to segmenting the convolution operation are special cases of block processing [3].

3.1.2 Circular Convolution

The circular convolution of two discrete-time signals, x(n) and h(n), is defined by the following equation:

$$y(n) = \sum_{m=0}^{N-1} h\langle n-m\rangle\, x(m) \qquad n=0,1,\ldots N-1 \quad (3.17)$$

In (3.17), the indices are evaluated modulo N with the symbol <n> representing the residue of n modulo N (see Section 2.1.3). The circular convolution in (3.17) is often written using the symbol "⊛".

$$y = h \circledast x \quad (3.18)$$

Circular convolution in (3.17) gives a length-N output when two length-N signals are convolved. Linear convolution, as defined in (3.1), gives a length-$(2N-1)$ output.

3.1.2.1 A Matrix View of Circular Convolution

The circular convolution of the signal x with the signal c may be viewed as a matrix multiplication as shown below for $N=5$.

$$\begin{bmatrix} y(0) \\ y(1) \\ y(2) \\ y(3) \\ y(4) \end{bmatrix} = \begin{bmatrix} c(0) & c(4) & c(3) & c(2) & c(1) \\ c(1) & c(0) & c(4) & c(3) & c(2) \\ c(2) & c(1) & c(0) & c(4) & c(3) \\ c(3) & c(2) & c(1) & c(0) & c(4) \\ c(4) & c(3) & c(2) & c(1) & c(0) \end{bmatrix} \begin{bmatrix} x(0) \\ x(1) \\ x(2) \\ x(3) \\ x(4) \end{bmatrix} \tag{3.19}$$

The matrix made up of circularly shifted versions of c is called a circulant matrix. This special structure allows very efficient calculation of a circular convolution. For example, by using the convolution property of the DFT (see Section 2.1.3), it is possible to compute a circular convolution using FFT algorithms.

3.1.2.2 Transform Property of Circular Convolution

If a length-N signal y is the result of the circular convolution of two length-N signals, h and x, then the DFT of y is equal to the product of the DFT of h and the DFT of x (see Section 2.1.3). In other words, if

$$y = h \circledast x \tag{3.20}$$

then

$$Y = H\,X \tag{3.21}$$

In contrast to the transform property of linear convolution, the z transform of the signal y in (3.20) is not equal to the product of the z transforms of h and x.

3.1.3 Linear Convolution Using Circular Convolution

When the linear convolution of a length-N signal $\bar{x}$ with a length-M signal $\bar{h}$ is performed, a length-$(M+N-1)$ signal y is produced. In order to calculate linear convolution using circular convolution, the appropriate number of zeros must be added to both $\bar{x}$ and $\bar{h}$. In particular, $M-1$ zeros are added to $\bar{x}$ to give $\bar{x}$, and $N-1$ zeros are added to $\bar{h}$ to give $\bar{h}$. The circular convolution of x and h, each of length-$(M+N-1)$, will then give the same length-$(M+N-1)$ output which would have been obtained with linear convolution. The matrix expression for circular convolution of two length-5 signals constructed from a length-2 $\bar{h}$ and a length-4 $\bar{x}$ is shown below.

$$\begin{bmatrix} y(0) \\ y(1) \\ y(2) \\ y(3) \\ y(4) \end{bmatrix} = \begin{bmatrix} h(0) & 0 & 0 & 0 & h(1) \\ h(1) & h(0) & 0 & 0 & 0 \\ 0 & h(1) & h(0) & 0 & 0 \\ 0 & 0 & h(1) & h(0) & 0 \\ 0 & 0 & 0 & h(1) & h(0) \end{bmatrix} \begin{bmatrix} x(0) \\ x(1) \\ x(2) \\ x(3) \\ 0 \end{bmatrix} \tag{3.22}$$

Notice that the circulation or wrapping around of the signal h in the upper right-hand corner of the circulant matrix caused no trouble, because it multiplies the zero which was appended to the length-4 signal $\bar{x}$ to make the length-5 signal, shown in (3.22).

By using a combination of segmentation, such as output overlap-add (described in Section 3.1.1.2), conversion of linear convolution to circular convolution, and the convolution property of the DFT (described in Section 2.1.3), it is possible to do convolution and FIR filtering of long signals very efficiently.

3.2 COMPUTING CONVOLUTION

In this section, algorithms for computing both linear and circular convolution are presented. The theory in Section 3.1 is used to compute linear convolution using FFT techniques and circular convolution. Minimum multiply algorithms for circular convolution, developed by Winograd, are also described.

3.2.1 Direct Calculation of Linear Convolution

The most straightforward way to compute convolution is to directly implement the definition (3.1) as illustrated by the FORTRAN statements in Figure 3-3. This figure illustrates the convolution of a length-N signal X(L) with a length-M signal H(L), giving a length-$(M+N-1)$ output Y(L). The M nonzero entries in the length-$(N+M-1)$ H array are located at index values of L from N to $N+M-1$, while the values of the H array for L from 1 to $N-1$ are zero.

```
      DO 20 J = 1 , N+M-1
         DO 10    L=1,N
         Y(J) = Y(J) + H(J-L+N)*X(L)
10       CONTINUE
20    CONTINUE
```

FIGURE 3-3. FORTRAN FOR DIRECT CONVOLUTION

There is a choice to be made between two direct methods of computing convolution. As shown in Figure 3-4, there are actually three different regions to consider if the number of program steps must be minimized.

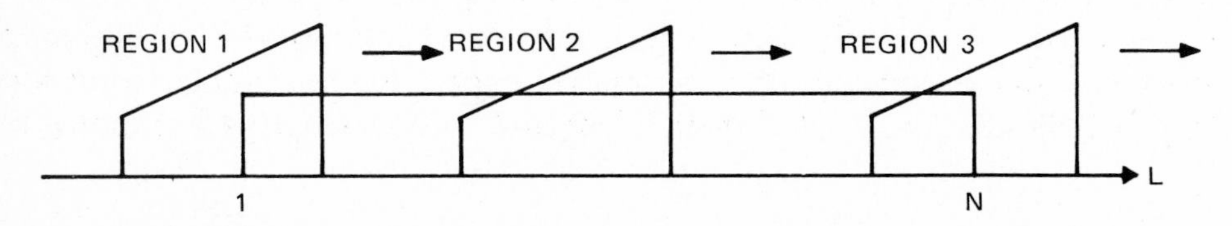

FIGURE 3-4. THREE REGIONS FOR DIRECT CONVOLUTION

In region 1, when $J < M+1$:

$$Y(J) = \sum_{L=1}^{J-1} H(J-L)X(L) \qquad J=1,M \tag{3.23}$$

In region 2, when $M+1 \leq \quad J < N+1$:

$$Y(J) = \sum_{L=J-M}^{J-1} H(J-L)X(L) \qquad J=M+1,N \tag{3.24}$$

In region 3, when $N < J$:

$$Y(J) = \sum_{L=J-M}^{N} H(J-L)\,X(L) \qquad J=N,\ N+M-1 \tag{3.25}$$

A program to make use of the three different summation limits indicated in (3.23 - 3.25) would require fewer multiply-adds than the simple statements in Figure 3-3 which require N multiply-adds for each output point. In Figure 3-3, even though the multiply involves a factor of zero, it is carried out. The more complicated program, however, may actually take longer to run because of the index testing which is necessary.

3.2.2 Direct Calculation of Circular Convolution

The direct calculation of length-N circular convolution, as defined in (3.9), requires an evaluation of the indices modulo N. If the computer does not have an efficient way to implement such a modulo N operation, then some extra index testing is required, as shown in Figure 3-5.

```
      DO 20 K=1, N
         I=K
         A=0
         DO 10 J=1, N
            A = A + X(J) * H(I)
            I=I-1
            IF (I.LE.0) I = I + N
10       CONTINUE
         Y(K) = A
20    CONTINUE
```

FIGURE 3-5. FORTRAN FOR CIRCULAR CONVOLUTION

While the simplified version of the modulo N operation on $I-1$ is an improvement, the testing of I (IF (I.LE.0))can be eliminated by storing a length-2N array for H, rather than the length-N array used in Figure 3-4. Linear convolution is then performed as in

Figure 3-3 using the extended H array. The length-2N array is simply formed by making a periodic repetition of H. In other words, the length-2N array would have entries of

$$H(1),H(2),\ldots, H(N),H(1),H(2),\ldots, H(N)$$

This is the alternate view of circular convolution as a linear convolution of periodic signals. The periodic extension of H(I) takes the place of modulo N evaluation of indices.

3.2.3 Minimum Multiply Convolution

Even though many algorithms had been proposed for the calculation of circular convolution, one could never be sure that there wasn't some other algorithm that would require fewer multiplications. Recently, some results from the mathematical theory of computational complexity have shown the absolute minimum number of multiplications which must be used for circular convolution. Not only is the minimum number of multiplications derived, but a constructive procedure is developed which aids the design of efficient algorithms.

3.2.3.1 Winograd's Circular Convolution

Winograd has shown how to calculate a circular convolution using the minimum number of multiplications. The Winograd convolution algorithms, while minimizing the number of multiplications, do not, however, simultaneously minimize the number of additions. For short circular convolutions (lengths up to 10 or 20), the number of additions is reasonable. Winograd [4] has shown that the number of multiplies which must be performed for a length-N circular convolution is $2N-1$ for prime N, and $2N-K$ for composite N where K is the number of relatively prime polynomial factors in the polynomial

$$P(z) = z^N - 1 \tag{3.26}$$

The minimum multiply convolutions first require additions of various elements of x and of h. Then, $2N-K$ multiplications are performed, followed by additions of elements in this intermediate product signal to give the length-N output. This procedure is illustrated for $N = 6$ [5] in Example 3-1.

Of course, a linear convolution of a length-M with a length-N signal could be computed with a length-$(M+N-1)$ circular convolution (described in Section 3.1.2). The length-6 circular convolution in Example 3-1 could be used to calculate linear convolution of a length-3 signal and a length-4 signal, or linear convolution of a length-2 signal with a length-5 signal, etc.

Example 3-1: Minimum Multiply Length-6 Circular Convolution

Since the polynomial z^6-1 has four irreducible factors, the minimum number of multiplications for a length-6 convolution is $2 \cdot 6 - 4$ or 8. The factored form of z^6-1 is

$$z^6 - 1 = (z+1)(z-1)(z^2+z+1)(z^2-z+1) \tag{3.27}$$

With the length-6 signals x and h represented as length-6 column vectors x, h, and y, the output is given by

$$y = C\,[\,Bh \otimes Ax\,] \tag{3.28}$$

The matrices A and B have eight rows and six columns while the matrix C has six rows and eight columns. The symbol $\otimes$ represents point-by-point multiplication of the two length-8 vectors. These are the only multiplications required for a length-6 circular convolution.

In this example,

$$A = \begin{bmatrix} 1 & -1 & 1 & -1 & 1 & -1 \\ 1 & 1 & 1 & 1 & 1 & 1 \\ 1 & -1 & 0 & 1 & -1 & 0 \\ 1 & 0 & -1 & 1 & 0 & -1 \\ 0 & 1 & -1 & 0 & 1 & -1 \\ 1 & 1 & 0 & -1 & -1 & 0 \\ 1 & 0 & -1 & -1 & 0 & 1 \\ 0 & 1 & 1 & 0 & -1 & -1 \end{bmatrix} = B \tag{3.29}$$

and

$$6C = \begin{bmatrix} 1 & 1 & 1 & 1 & -2 & 1 & 1 & -2 \\ -1 & 1 & -2 & 1 & 1 & 2 & -1 & -1 \\ 1 & 1 & 1 & -2 & 1 & 1 & -2 & 1 \\ -1 & 1 & 1 & 1 & -2 & -1 & -1 & 2 \\ 1 & 1 & -2 & 1 & 1 & -2 & 1 & 1 \\ -1 & 1 & 1 & -2 & 1 & -1 & 2 & 1 \end{bmatrix} \tag{3.30}$$

3.2.4 Fast Convolution Using the FFT

The procedure of using the convolution property of the DFT (see Section 2.1.3), combined with input segmentation into blocks of length-N is called "fast convolution", or "high-speed convolution" [6]. Fast convolution is indeed the fastest way to implement convolution for long input signals, medium-length filters, and long multiply-add times.

As described in Section 3.2.1, a length-$(M+B-1)$ circular convolution can be used to compute the linear convolution of a length-M signal (considered here to be the filter) and a length-B signal (considered to be the input block). At first, one might think that the block length, B, should be chosen to be about the same length as the filter length, M. It is possible, however, to use a longer circular convolution to advantage. If the input signal is segmented into length-B non-overlapping blocks and the output overlap-add method of (3.10-3.13) is used, a circular convolution of length $L = M+B-1$ would be calculated for each of the input segments. If the complete input signal were segmented into K length-B blocks, the time to compute a fast convolution would be

$$T_{fast} = T_{fft} + 2KT_{fft} + KLT_{aux} \tag{3.31}$$

where T_{fft} is the time required for a length-L FFT and T_{aux} is the time required for the auxiliary calculations associated with the data sectioning in the overlap-add method, as well as the time required for the point-by-point frequency domain multiplications. The first term on the right side of (3.31) represents the length-L FFT of the filter impulse response, augmented with zeros, which is only calculated once. The next term represents the forward transforms of the data blocks and the inverse transform of the product of the data transforms and the filter transform. The third term represents the K point-by-point multiplications of transform values and the auxiliary overlap-add calculations.

The block-length B should be chosen to minimize the time T_{fast} for the particular hardware being used for the signal processing. When using a radix-2 FFT on an IBM 7094, Stockham [6] found that a length-128 FFT should be used for a length-24 filter, while a length-256 FFT should be used for a length-48 filter. When he compared actual computing times, he discovered that it was faster to do the length-24 convolution directly. For the length-48 convolution, the fast convolution method was 50 percent faster than direct convolution.

When considering whether to use direct or fast convolution for the TMS32010 processor, a calculation similar to that in (3.31) must be made for the data and filter length involved. The crossover from direct convolution to fast convolution will occur for longer filters than those discussed above because of the high-speed multiplier used on the TMS32010 and the penalty in time which arises from the auxiliary calculations for the overlap-add data blocking. Since buffer memory is required to store 4N complex numbers, where N is the block length, fast convolution is probably not a practical alternative for the TMS32010.

3.2.5 Multidimensional Index Mapping

The multidimensional index mapping, described in Chapter 2 for breaking up a large one-dimensional DFT into several small transforms, can be used directly on the circular convolution defined in (3.9) [7]. To illustrate the theory involved, a mapping from one to two dimensions is described. The original one-dimensional circular convolution is

$$y(n) = \sum_{m=0}^{M-1} h\langle n-m\rangle\, x(m) \tag{3.32}$$

With a Chinese Remainder Theorem index mapping, (described in Chapter 2), $n \to (n_1, n_2)$ and the single sum in (3.32) becomes a double summation.

$$y(n_1,n_2) = \sum_{m_1} \sum_{m_2} h\langle n_1 - m_1, n_2 - m_2\rangle\, x(m_1,m_2) \tag{3.33}$$

In this two-dimensional convolution, no operations are saved over the one-dimensional convolution in (3.32), but the length-($N = M_1M_2$) convolution is computed using shorter length-M_1 and length-M_2 circular convolutions. These could be efficiently computed using the Winograd minimum multiply convolution described in Section 3.2.3.1.

One possible application of this multidimensional mapping allows the double-length circular convolution to be computed by doing four convolutions of the original length. This flexibility may be very useful when programming the TMS32010 with its limited data memory.

After mapping the input of length-2L and the impulse response into two-dimensional arrays, x(m,n) and h(m,n), the four convolutions which must be computed are given by

$$y(m,n) = \sum_{l=0}^{1} \sum_{k=0}^{L} h\langle m-l, n-k\rangle \, x(l,k) \, . \tag{3.34}$$

Writing out these terms gives

$$y(0,n) = \sum_{k=0}^{L-1} h\langle 0, n-k\rangle \, x(0,k) + \sum_{k=0}^{L-1} h\langle -1, n-k\rangle \, x(1,k) \tag{3.35}$$

$$y(1,n) = \sum_{k=0}^{L-1} h\langle 1, n-k\rangle \, x(0,k) + \sum_{k=0}^{L-1} h\langle 0, n-k\rangle \, x(1,k) \tag{3.36}$$

The final output signal of length-2L is found by using the inverse index mapping described in Section 2.2.7. The four convolutions indicated in (3.34) may be reduced to only three convolutions by using a method similar to that for computing a complex multiplication that uses only three real multiplications [4].

3.3 REFERENCES

1. Rabiner, L.R. and Gold, B., THEORY AND APPLICATION OF DIGITAL SIGNAL PROCESSING. Englewood Cliffs, NJ: Prentice-Hall, Inc., 1975.

2. Oppenheim, A.V. and Schafer, R.W., DIGITAL SIGNAL PROCESSING. Englewood Cliffs, NJ: Prentice-Hall, Inc., 1975.

3. Burrus, C.S., "Block Realization of Digital Filters," IEEE TRANS. ON ASSP, Vol. 20, 1972, 213-218.

4. Winograd, S., "Some Bilinear Forms Whose Multiplicative Complexity Depends on the Field of Constants,"IBM RESEARCH REPORT RC 5669, October 10, 1975.

5. Kolba, D.P. and Parks, T.W., "A Prime Factor FFT Algorithm Using High-Speed Convolution," IEEE TRANS. ON ASSP, Vol. 25, 1977, 281-294.

6. Stockham, T.G.,"High-Speed Convolution and Correlation," reprinted in DIGITAL SIGNAL PROCESSING (edited by L.R. Rabiner and C.M. Rader). New York, NY: IEEE PRESS, 1972.

7. Agarwal, R. and Cooley, J.W., "Algorithms for Digital Convolution," IEEE TRANS. ON ASSP, Vol. 25, 1977, 392-410.

Chapter 4

FORTRAN PROGRAMS FOR THE DFT AND CONVOLUTION

4.1 INTRODUCTION

In this chapter, the theory and ideas of Chapters 2 and 3 are made specific in computer programs. First, FORTRAN programs are given and described in a way that helps in the implementation of these efficient algorithms. The use of a high-level language makes programs much easier to write, debug, modify, maintain, and explain. After the more useful algorithms are described in FORTRAN code, those that look promising for implementation on the TMS32010 are essentially hand-compiled to TMS32010 assembly language code and presented in Chapter 5.

The FORTRAN programs have been tested and debugged on mainframe computers, and are written to be as efficient as possible without becoming difficult to read and without using compiler or machine-dependent features. They should be used in conjuction with the theory from Chapter 2 to develop optimal programs for a particular application. Likewise, the TMS32010 code should be used as a starting point for the implementation of an algorithm from the theory or from a high-level language program.

All of the nineteen FORTRAN programs are written as subroutines to be called by a main program that supplies the input data, the array sizes, and other parameters to be passed to the subroutine. Two sample calling programs are given at the end of the program collection in the form of testing programs used to check the correctness of the DFT subroutines. The input is passed to the subroutine with the real part in the X array and the imaginary part in the Y array; the real part of the output is put in the A array and the imaginary part in the B array for algorithms that are not in-place. All are listings of tested, working programs.

4.2 FORTRAN PROGRAMS

The first program, listed in Table 4-1 as Program 1, is a subroutine implementing the direct calculation of the DFT from Equations (2.28) and (2.31). This program is similar to the program in Figure 2-3. In an actual application, the calculation of the trig table by the DO 10 loop would be done as an initializing process, not as part of the DFT calculation. It is not possible to do "in-place" calculations with this approach; therefore, separate input and output arrays must exist.

Program 2 is an implementation of a first-order Goertzel algorithm from Equation (2.38) and an extension of Figure 2-6. More efficient forms are given in the second-order

Goertzel algorithms in Programs 3 and 4 which are implementations of (2.43) and (2.47) and extensions of Figure 2-7. In Program 3, the data is used in reverse order, and in Program 4, it is used in forward order. However, both have the same arithmetic efficiency.

The basic radix-2 Cooley-Tukey decimation-in-frequency FFT algorithm, described in Sections 2.2.9-11, is implemented in Program 5. This program takes complex input data in arrays X and Y and calculates the DFT in-place, i.e., writes the output back into the X and Y arrays over the input data, which is destroyed. This program uses the index map in (2.86) and has M stages of the form in (2.87) as illustrated in Figures 2-10 and 2-11. The length of the data must be $N = 2^M$, and the outer DO 10 loop steps through the M stages. The twiddle factors are evaluated inside the DO 20 loop by the cosine and sine functions. The actual DFT evaluation is done in the innermost loop, the DO 30 loop. The first statement calculates an address offset for the butterfly. The next four statements calculate the length-2 DFT butterfly from (2.92) and (2.94). The last two statements in the loop are the twiddle-factor complex multiplication using four real multiplies and two real adds, as shown in (2.89) and (2.96) and illustrated in Figures 2-11 and 2-12.

The in-place output of the basic FFT algorithm is in a scrambled order, as explained in Section 2.2.8. The last part of Program 5 uses an in-place unscrambler called a Digit Reverse Counter. As the index I steps normally from one to N-1, the index J steps in bit-reverse order, and the output is reordered with these two indices. The details of this counter can be found in [1] or [2]. There are many applications where the DFT does not have to be used in proper order, and the unscrambler can be omitted. When the FFT is being used for high-speed convolution (described in Chapter 3), the coefficients can be used in scrambled order. The inverse FFT algorithm could be the decimation-in-time version (see Section 2.2.11) which normally has the unscrambler before the main FFT section, and no unscrambling need be done in the transform or the inverse transform. The unscrambler is the same for any radix-2 FFT; therefore, it will not be repeated in the other programs.

One of the time-consuming operations in Program 5 is the calculation of the cosine and sine functions. This is eliminated in Program 6 by the use of a precomputed table generated by Program 7. Because of the structure of the FFT, no residue reduction is necessary on the address calculation of the variable IA. This makes the use of tables even easier than for the direct or the Goertzel algorithms. A compromise method for generating the twiddle factors is used in Program 8, where different cosines and sines are calculated in the outer loop for each stage and are updated as needed within each stage by the statements between labels 30 and 20. This method uses almost as little memory as Program 5 but runs almost as fast as Program 6. Even the small number of trigometric function evaluations in Program 8 could be eliminated by use of a small table. This updating approach is somewhat similar to the generation of trigometric functions that takes place in the Goertzel algorithm. The disadvantage of this approach is some error accumulation in the use of an update rather than a regeneration process.

The removal of certain multiplications by twiddle factors that equal unity is implemented in Program 9 by writing a special butterfly with no built-in twiddle factors. This special butterfly is in the DO 1 loop. All other parts are the same as in Program 6. The use of the update method in Program 8 could also easily be incorporated here. The only disadvantage is slightly longer code. In certain cases with the TMS32010 processor, the test may take more time than the multiplications. Removal of multiplies by unity or j also reduces the quantization error [18].

Program 10 removes still more unnecessary multiplications by removing those by j and reducing those described in (2.101). This is done by having a third butterfly to calculate that special case, which is trapped by the IF statement just after the DO 20 statement. Again, the reduced multiplication may not offset the time required by the IF statement. In this case, the test statement is costly, because it is inside the second level of looping.

One of the most effective means of improving efficiency is to use a larger radix of 4, 8, or even 16. Program 11 is a basic radix-4 FFT, similar in structure to Program 5, but using the radix-4 butterfly of (2.99), a radix-4 digit-reverse counter for unscrambling, and a sequence length of N = 4**M. As for the radix-2 cases, the radix-4 unscrambler will not be repeated in the other radix-4 FFT programs where it may or may not be needed according to the application.

Program 12 incorporates the same principles as in Programs 6 and 10 for table lookup of the twiddle factors coupled with three butterflies to remove all multiplications by 1 and j, and to reduce some of those discussed in (2.101). This is a very efficient program on a general-purpose computer but it may or may not be appropriate for the TMS32010.

Programs 13 and 14 are modifications of Program 12. Program 13 uses a reduced number of temporary variables following the ideas of Morris [5], and Program 14 uses the three-multiplication algorithm for complex multiplication as described in (2.102) and (2.103). Both of these modifications must be checked carefully to see if they are warranted on the TMS32010.

Some improvement is possible by using a radix-8 FFT. Program 15 gives the basic radix-8 FFT with the unscrambler in the same form as in Programs 5 and 11 for radix-2 and radix-4 FFTs. Program 16 uses two butterflies and table lookup. There is little advantage to using three butterflies with the radix-8 form. Table 2-4 should be consulted for the numbers of multiplies and adds for the various forms of the FFT. The radix-8 FFT has a few more adds than the radix-4 FFT, but the total number of multiplies and adds decreases.

Although not presented separately in this book, a radix-16 FFT could easily be written by taking the length-16 butterfly from the length-16 module of the PFA in Program 17 and using it with the structure of any of the other radices in this section. The savings in arithmetic as shown in Table 2-4 indicates very little potential improvement over the radix-8 or even the radix-4.

There are several FORTRAN programs in [4],[5], and [10] that should be considered in addition to those listed in this chapter. Morris presents a type of in-line FORTRAN program for a radix-4 FFT in [5] which is rather long but fast.

A very efficient realization of the prime-factor algorithm (PFA) is presented in Program 17. This follows the theory from Section 2.2.10 and is an expansion of Figure 2-16. Efficient modules for lengths of 2, 3, 4, 5, 7, 8, 9, and 16 are part of the program as is an unscrambler. The PFA algorithm itself is in-place but the unscrambler is not, thus resulting in the total program not being in-place. This requires separate output arrays A and B if an in-order result is required. The number of multiplier coefficients entered as DATA values in this program is much smaller than the size of the tables required by the table lookup forms of the FFT or direct structures in Programs 3, 13, 20, 28, etc.

The possible DFT lengths and required number of operations are given in Table 2-6. In a practical application, only the modules and multiplier coefficients needed for a desired total DFT length would be used. If other lengths are needed, modules for 11, 13, 17, 19, and 25 can be found in [10].

If an in-place algorithm is needed because of memory limitations and if the DFT must be in proper order, there are three ways to achieve this. The first is given in Program 18 which uses a separate output index pointer, IP, to perform the unscrambling during the output of each module. This requires little overhead on a general-purpose computer, but more on the TMS32010. Setting up this output pointer is done by the DO 15 loop, which is not present in Program 17. The output portion of each module is also modified to use the output pointer IP, rather than I. Only four modules are given in this program in order to avoid repetition. The others can be added by modifying the appropriate modules from Program 17.

The second method applies if only one length DFT is to be calculated. Then it is possible to avoid both the unscrambler of Program 17 and the separate output index pointer of Program 18. The permutation on the output index used in Program 18 in the DO 15 loop can be precomputed and built into the output parts of the modules. This is illustrated in Program 19 for a length-15 DFT. The form of the index loops is exactly the same as in Program 17, but there is no unscrambler. Note the order of the outputs in each short module has been changed from that in Program 17 by exactly what would have been done by the output pointer in Program 18. This could be done for any total length, but the output modification would have to be calculated for that particular length.

The third method achieves the same results as in Program 19 by changing the values of the multiplier coefficients rather than the output locations. Calculating the required coefficient values is somewhat complicated and therefore will not be covered here but can be found in [14].

A new type of Cooley-Tukey FFT, called the Split-Radix FFT, is given in Program 20. The theory of this approach is given in [19] which uses a mixture of the radix-2 and radix-4 index map. The FORTRAN program illustrates a basic one-butterfly structure. Modifications which add special butterflies to remove multiplies by unity and reduce those by the eighth root of W give a total operation count that is the lowest known. A complex four-butterfly split-radix FFT using the three-multiply complex multiply algorithm in (2.105) requires $N(M-1) + 4$ real multiplications and $3N(M-1) + 4$ real additions, which are fewer than the five-butterfly radix-16 FFT and the same as obtained for the shorter lengths by Winograd's methods in Table 2-5 and the same as reported in [20]. The output order is scrambled in a more complicated way than for the constant radix FFT, and the indexing is slightly more complicated; however, the total operation count seems to be the smallest for a power-of-two Cooley-Tukey FFT algorithm. This program is worth further development.

Although a FORTRAN program for the WFTA is not given in this chapter, one could be derived by extending the basic form given in Figure 2-21 or in [4] or [8]. The resulting programs are fairly long with a large multiplier coefficient table, and the calculations are not in-place. Table 2-8 shows that the WFTA has fewer multiplications than the FFT or PFA but has more additions. There are some applications where this tradeoff is an advantage but with systems having a very fast multiplication like the TMS 32010, there is no advantage and there are the previously mentioned disadvantages.

When a particular algorithm has been chosen for consideration, it should first be developed from the examples given here, tested and debugged in high-level language form, hand-translated into assembly language, and finally tested and debugged again. Two test programs are given in FORTRAN by Programs 21 and 22. The first program applies a complex geometric sequence as a test input, because it has a known DFT that can be compared to the result found by the algorithm being tested. This is a fairly complete test, but it gives little clue as to how to debug an incorrect program. The second test program is more helpful in that respect. It applies an impulse input, and as shown in Section 2.1, the output should be a sampled sinusoid. By examining the calculated output from various impulse inputs, the errors can usually be found. Another informative test is to use a sampled sinusoid as an input giving a single impulse as a DFT. Even after the tested program is corrected in form, the small residual values at the wrong frequencies are a measure of the effects of quantization error due to the finite word length [18].

4.3 REFERENCES

The references for this chapter are the same as for Chapter 2.

4.4 SAMPLE FORTRAN PROGRAMS

The following pages present sample FORTRAN programs for the DFT. Table 4-1 lists the programs that are presented.

TABLE 4-1. FORTRAN DFT PROGRAM INDEX

```
C
C    DFT SUBROUTINE WITH TABLE LOOK-UP
C
C    C. S. BURRUS,  SEPT 1983
C
C---------------------------------------------
C
      SUBROUTINE DFT(X,Y,A,B,N)
      REAL X(1), Y(1), A(1), B(1), C(60), S(60)
C
      Q = 6.283185307179586/N
      DO 10 J = 1, N
          C(J) = COS(Q*(J-1))
          S(J) = SIN(Q*(J-1))
  10  CONTINUE
C
      DO 20 J = 1, N
          AT = X(1)
          BT = Y(1)
          K  = 1
          DO 30 I = 2, N
              K  = K + J - 1
              IF (K.GT.N)  K = K - N
              AT = AT + C(K)*X(I) + S(K)*Y(I)
              BT = BT + C(K)*Y(I) - S(K)*X(I)
   30     CONTINUE
          A(J) = AT
          B(J) = BT
   20 CONTINUE
C
      RETURN
      END
```

```
C
C   GOERTZEL'S  DFT  ALGORITHM
C          FIRST ORDER
C   C. S. BURRUS,    SEPT 1983
C
C-------------------------------------------------
C
        SUBROUTINE DFT(X,Y,A,B,N)
        REAL X(1), Y(1), A(1), B(1)
C
        Q = 6.283185307179586/N
        DO 20 J=1, N
            C  = COS(Q*(J-1))
            S  = SIN(Q*(J-1))
            AT = X(N)
            BT = Y(N)
            DO 30 I = 1, N-2
                T  = C*AT + S*BT + X(N-I)
                BT = C*BT - S*AT + Y(N-I)
                AT = T
   30       CONTINUE
            A(J) = C*AT + S*BT + X(1)
            B(J) = C*BT - S*AT + Y(1)
   20   CONTINUE
C
        RETURN
        END
```

```
C
C   GOERTZEL'S  DFT  ALGORITHM
C       SECOND ORDER
C       REVERSE ORDER INPUT
C   C. S. BURRUS,   SEPT 1983
C
C---------------------------------------------
C
      SUBROUTINE DFT(X,Y,A,B,N)
      REAL X(1), Y(1), A(1), B(1)
C
      Q = 6.283185307179586/N
      DO 20 J = 1, N
          C  = COS(Q*(J-1))
          S  = SIN(Q*(J-1))
          CC = 2*C
          A2 = 0
          B2 = 0
          A1 = X(N)
          B1 = Y(N)
          DO 30 I = 1, N-2
              T  = A1
              A1 = CC*A1 - A2 + X(N-I)
              A2 = T
              T  = B1
              B1 = CC*B1 - B2 + Y(N-I)
              B2 = T
   30     CONTINUE
          A(J) = C*A1 - A2 + S*B1 + X(1)
          B(J) = C*B1 - B2 - S*A1 + Y(1)
   20 CONTINUE
C
      RETURN
      END
```

```
C
C   GOERTZEL'S  DFT  ALGORITHM
C
C   C. S. BURRUS,    SEPT 1983
C
C---------------------------------------------
C
      SUBROUTINE DFT(X,Y,A,B,N)
      REAL X(1), Y(1), A(1), B(1)
C
      Q = 6.283185307179586/N
      DO 20 J = 1, N
          C  = COS(Q*(J-1))
          S  = SIN(Q*(J-1))
          CC = 2*C
          A2 = 0
          B2 = 0
          A1 = X(1)
          B1 = Y(1)
          DO 30 I = 2, N
              T  = A1
              A1 = CC*A1 - A2 + X(I)
              A2 = T
              T  = B1
              B1 = CC*B1 - B2 + Y(I)
              B2 = T
   30     CONTINUE
          A(J) = C*A1 - A2 - S*B1
          B(J) = C*B1 - B2 + S*A1
   20 CONTINUE
C
      RETURN
      END
```

```
C
C       A COOLEY-TUKEY RADIX-2, DIF  FFT PROGRAM
C       COMPLEX INPUT DATA IN ARRAYS X AND Y
C
C          C. S. BURRUS, RICE UNIVERSITY, SEPT 1983
C
C---------------------------------------------------------------------
C
        SUBROUTINE FFT (X,Y,N,M)
        REAL X(1), Y(1)
C
C--------------MAIN FFT LOOPS-----------------------------
C
        N2 = N
        DO 10 K = 1, M
            N1 = N2
            N2 = N2/2
            E  = 6.283185307179586/N1
            A  = 0
            DO 20 J = 1, N2
                C = COS (A)
                S = SIN (A)
                A = J*E
                DO 30 I = J, N, N1
                    L = I + N2
                    XT   = X(I) - X(L)
                    X(I) = X(I) + X(L)
                    YT   = Y(I) - Y(L)
                    Y(I) = Y(I) + Y(L)
                    X(L) = C*XT + S*YT
                    Y(L) = C*YT - S*XT
   30           CONTINUE
   20       CONTINUE
   10   CONTINUE
C
C------------DIGIT REVERSE COUNTER-----------------
C
  100   J = 1
        N1 = N - 1
        DO 104 I=1, N1
            IF (I.GE.J) GOXTO 101
            XT = X(J)
            X(J) = X(I)
            X(I) = XT
            XT   = Y(J)
            Y(J) = Y(I)
            Y(I) = XT
  101       K = N/2
  102       IF (K.GE.J) GOTO 103
                J = J - K
                K = K/2
                GOTO 102
  103       J = J + K
  104   CONTINUE
        RETURN
        END
```

```
C
C       A COOLEY-TUKEY RADIX-2, DIF  FFT PROGRAM
C       COMPLEX INPUT DATA IN ARRAYS X AND Y
C       TABLE LOOK-UP OF W VALUES
C
C          C. S. BURRUS, RICE UNIVERSITY, SEPT 1983
C
C---------------------------------------------------------------------
C
        SUBROUTINE FFT (X,Y,N,M,WR,WI)
        REAL X(1), Y(1), WR(1), WI(1)
C
C--------------MAIN FFT LOOPS-----------------------------------
C
        N2 = N
        DO 10 K = 1, M
            N1 = N2
            N2 = N2/2
            IE = N/N1
            IA = 1
            DO 20 J = 1, N2
                C = WR(IA)
                S = WI(IA)
                IA =IA + IE
                DO 30 I = J, N, N1
                    L = I + N2
                    XT   = X(I) - X(L)
                    X(I) = X(I) + X(L)
                    YT   = Y(I) - Y(L)
                    Y(I) = Y(I) + Y(L)
                    X(L) = C*XT + S*YT
                    Y(L) = C*YT - S*XT
   30           CONTINUE
   20       CONTINUE
   10   CONTINUE
C
C------------DIGIT REVERSE COUNTER-----------------
C
  100   J = 1
        N1 = N - 1
        DO 104 I=1, N1
            IF (I.GE.J) GOTO 101
            XT = X(J)
            X(J) = X(I)
            X(I) = XT
            XT   = Y(J)
            Y(J) = Y(I)
            Y(I) = XT
  101       K = N/2
  102       IF (K.GE.J) GOTO 103
                J = J - K
                K = K/2
                GOTO 102
  103       J = J + K
  104   CONTINUE
        RETURN
        END
```

```
C
C       INITIALIZE SINE AND COSINE TABLES
C       FOR FFTs WITH TABLE LOOKUP OF TFs
C
C       C.S. BURRUS,    RICE UNIVERSITY
C
C               DEC 1983
C
        SUBROUTINE INI (N, WR, WI)
        REAL WR(1),WI(1)
C
        P = 6.283185307179586/N
        DO 10 K=1, N
           A = (K-1)*P
           WR(K) = COS(A)
           WI(K) = SIN(A)
   10   CONTINUE
        RETURN
        END
```

```
C
C       A COOLEY-TUKEY RADIX-2, DIF  FFT PROGRAM
C       COMPLEX INPUT DATA IN ARRAYS X AND Y
C       TWIDDLE FACTOR  W  UP-DATE
C
C          C. S. BURRUS, RICE UNIVERSITY, SEPT 1983
C
C---------------------------------------------------------------
C
        SUBROUTINE FFT (X,Y,N,M)
        REAL X(1), Y(1)
C
C--------------MAIN FFT LOOPS-----------------------------
C
        N2 = N
        DO 10 K = 1, M
            N1 = N2
            N2 = N2/2
            E  = 6.283185307179586/N1
            C = 1
            S = 0
            C1 = COS (E)
            S1 = SIN (E)
            DO 20 J = 1, N2
                DO 30 I = J, N, N1
                    L = I + N2
                    XT   = X(I) - X(L)
                    X(I) = X(I) + X(L)
                    YT   = Y(I) - Y(L)
                    Y(I) = Y(I) + Y(L)
                    X(L) = C*XT + S*YT
                    Y(L) = C*YT - S*XT
   30           CONTINUE
            T = C
            C = C*C1 - S*S1
            S = T*S1 + S*C1
   20       CONTINUE
   10   CONTINUE
C
C------------DIGIT REVERSE COUNTER-----------------
C
  100   J = 1
        N1 = N - 1
        DO 104 I=1, N1
            IF (I.GE.J) GOTO 101
            XT = X(J)
            X(J) = X(I)
            X(I) = XT
            XT   = Y(J)
            Y(J) = Y(I)
            Y(I) = XT
  101       K = N/2
  102       IF (K.GE.J) GOTO 103
                J = J - K
                K = K/2
                GOTO 102
  103       J = J + K
  104   CONTINUE
        RETURN
        END
```

```
C
C        A COOLEY-TUKEY RADIX-2, DIF  FFT PROGRAM
C        TWO-BFs, MULTIPLICATIONS BY  1  ARE REMOVED
C        COMPLEX INPUT DATA IN ARRAYS X AND Y
C        TABLE LOOK-UP OF W VALUES
C
C           C. S. BURRUS, RICE UNIVERSITY, SEPT 1983
C
C---------------------------------------------------------------
C
         SUBROUTINE FFT (X,Y,N,M,WR,WI)
         REAL X(1), Y(1), WR(1), WI(1)
C
C--------------MAIN FFT LOOPS-----------------------------------
C
         N2 = N
         DO 10 K = 1, M
             N1 = N2
             N2 = N2/2
             DO 1 I = 1, N, N1
                 L = I + N2
                 T    = X(I) - X(L)
                 X(I) = X(I) + X(L)
                 X(L) = T
                 T    = Y(I) - Y(L)
                 Y(I) = Y(I) + Y(L)
                 Y(L) = T
   1         CONTINUE
             IF (K.EQ.M) GOTO 10
             IE  = N/N1
             IA  = 1
             DO 20 J = 2, N2
                 IA = IA + IE
                 C = WR(IA)
                 S = WI(IA)
                 DO 30 I = J, N, N1
                     L = I + N2
                     T    = X(I) - X(L)
                     X(I) = X(I) + X(L)
                     TY   = Y(I) - Y(L)
                     Y(I) = Y(I) + Y(L)
                     X(L) = C*T + S*TY
                     Y(L) = C*TY - S*T
  30             CONTINUE
  20         CONTINUE
  10     CONTINUE
C
C------------DIGIT REVERSE COUNTER-----------------
C               SAME AS PROGRAM 5
         RETURN
         END
```

```
C
C       A COOLEY-TUKEY RADIX 2, DIF  FFT PROGRAM
C       THREE-BF, MULT BY  1  AND  J  ARE REMOVED
C       COMPLEX INPUT DATA IN ARRAYS X AND Y
C       TABLE LOOK-UP OF W VALUES
C
C          C. S. BURRUS, RICE UNIVERSITY, SEPT 1983
C
C---------------------------------------------------------
C
        SUBROUTINE FFT (X,Y,N,M,WR,WI)
        REAL X(1), Y(1), WR(1), WI(1)
C
C--------------MAIN FFT LOOPS-----------------------------
C
        N2 = N
        DO 10 K = 1, M
            N1 = N2
            N2 = N2/2
            JT = N2/2 + 1
            DO 1 I = 1, N, N1
                L = I + N2
                T    = X(I) - X(L)
                X(I) = X(I) + X(L)
                X(L) = T
                T    = Y(I) - Y(L)
                Y(I) = Y(I) + Y(L)
                Y(L) = T
   1        CONTINUE
            IF (K.EQ.M) GOTO 10
            IE   = N/N1
            IA   = 1
            DO 20 J = 2, N2
                IA = IA + IE
                IF (J.EQ.JT) GOTO 50
                C = WR(IA)
                S = WI(IA)
                DO 30 I = J, N, N1
                    L = I + N2
                    T    = X(I) - X(L)
                    X(I) = X(I) + X(L)
                    TY   = Y(I) - Y(L)
                    Y(I) = Y(I) + Y(L)
                    X(L) = C*T + S*TY
                    Y(L) = C*TY - S*T
  30            CONTINUE
                GOTO 25
  50            DO 40 I = J, N, N1
                    L = I + N2
                    T    = X(I) - X(L)
                    X(I) = X(I) + X(L)
                    TY   = Y(I) - Y(L)
                    Y(I) = Y(I) + Y(L)
                    X(L) = TY
                    Y(L) =-T
  40            CONTINUE
  25        A = J*E
  20        CONTINUE
  10    CONTINUE
C
C------------DIGIT REVERSE COUNTER------------------
C              SAME AS PROGRAM 1
        RETURN
        END
```

```
C       A COOLEY-TUKEY RADIX-4 DIF  FFT PROGRAM
C       COMPLEX INPUT DATA IN ARRAYS X AND Y
C       LENGTH IS  N = 4 ** M
C
C         C. S. BURRUS, RICE UNIVERSITY, SEPT 1983
C
C---------------------------------------------------------------
C
        SUBROUTINE  FFT4 (X,Y,N,M)
        REAL X(1), Y(1)
C-------------MAIN FFT LOOPS-----------------------------
        N2 = N
        DO 10 K = 1, M
            N1 = N2
            N2 = N2/4
            E = 6.283185307179586/N1
            A = 0
            DO 20 J=1, N2
                B    = A + A
                C    = A + B
                CO1  = COS(A)
                CO2  = COS(B)
                CO3  = COS(C)
                SI1  = SIN(A)
                SI2  = SIN(B)
                SI3  = SIN(C)
                A    = J*E
                DO 30 I=J, N, N1
                    I1 = I  + N2
                    I2 = I1 + N2
                    I3 = I2 + N2
                    R1 = X(I ) + X(I2)
                    R3 = X(I ) - X(I2)
                    S1 = Y(I ) + Y(I2)
                    S3 = Y(I ) - Y(I2)
                    R2 = X(I1) + X(I3)
                    R4 = X(I1) - X(I3)
                    S2 = Y(I1) + Y(I3)
                    S4 = Y(I1) - Y(I3)
                    X(I) = R1 + R2
                    R2   = R1 - R2
                    R1   = R3 - S4
                    R3   = R3 + S4
                    Y(I) = S1 + S2
                    S2   = S1 - S2
                    S1   = S3 + R4
                    S3   = S3 - R4
                    X(I1) = CO1*R3 + SI1*S3
                    Y(I1) = CO1*S3 - SI1*R3
                    X(I2) = CO2*R2 + SI2*S2
                    Y(I2) = CO2*S2 - SI2*R2
                    X(I3) = CO3*R1 + SI3*S1
                    Y(I3) = CO3*S1 - SI3*R1
  30            CONTINUE
  20        CONTINUE
  10    CONTINUE
```

```
C-----------DIGIT REVERSE COUNTER----------
  100   J = 1
        N1 = N - 1
        DO 104 I = 1, N1
            IF (I.GE.J) GOTO 101
            R1   = X(J)
            X(J) = X(I)
            X(I) = R1
            R1   = Y(J)
            Y(J) = Y(I)
            Y(I) = R1
 101        K = N/4
 102        IF (K*3.GE.J) GOTO 103
                J = J - K*3
                K = K/4
                GOTO 102
 103        J = J + K
 104    CONTINUE
        RETURN
        END
```

```
C
C        A COOLEY-TUKEY RADIX-4 DIF  FFT PROGRAM
C        THREE BF, MULTIPLICATIONS BY  1, J, ETC. ARE REMOVED
C        COMPLEX INPUT DATA IN ARRAYS X AND Y
C        LENGTH IS  N = 4 ** M
C        TABLE LOOKUP OF W VALUES
C
C          C. S. BURRUS, RICE UNIVERSITY,  SEPT 1983
C
C---------------------------------------------------------------
C
         SUBROUTINE  FFT4 (X,Y,N,M,WR,WI)
         REAL X(1), Y(1), WR(1), WI(1)
         DATA C21 / 0.707106778 /
C
C--------------MAIN FFT LOOPS-----------------------------------
C
         N2 = N
         DO 10 K = 1, M
             N1 = N2
             N2 = N2/4
             JT = N2/2 + 1
C---------------SPECIAL BUTTERFLY FOR W = 1--------------
             DO 1 I = 1, N, N1
                 I1 = I  + N2
                 I2 = I1 + N2
                 I3 = I2 + N2
                     R1 = X(I ) + X(I2)
                     R3 = X(I ) - X(I2)
                     S1 = Y(I ) + Y(I2)
                     S3 = Y(I ) - Y(I2)
                     R2 = X(I1) + X(I3)
                     R4 = X(I1) - X(I3)
                     S2 = Y(I1) + Y(I3)
                     S4 = Y(I1) - Y(I3)
                     X(I) = R1 + R2
                     X(I2)= R1 - R2
                     X(I3)= R3 - S4
                     X(I1)= R3 + S4
                     Y(I) = S1 + S2
                     Y(I2)= S1 - S2
                     Y(I3)= S3 + R4
                     Y(I1)= S3 - R4
   1         CONTINUE
             IF (K.EQ.M) GOTO 10
             IE = N/N1
             IA1 = 1
C--------------GENERAL BUTTERFLY-----------------
             DO 20 J = 2, N2
                 IA1  = IA1 + IE
                 IF (J.EQ.JT) GOTO 50
                 IA2  = IA1 + IA1 - 1
                 IA3  = IA2 + IA1 - 1
                 CO1  = WR(IA1)
                 CO2  = WR(IA2)
                 CO3  = WR(IA3)
                 SI1  = WI(IA1)
                 SI2  = WI(IA2)
                 SI3  = WI(IA3)
```

```
C---------------BUTTERFLIES WITH SAME W---------------
                DO 30 I = J, N, N1
                    I1 = I  + N2
                    I2 = I1 + N2
                    I3 = I2 + N2
                    R1 = X(I ) + X(I2)
                    R3 = X(I ) - X(I2)
                    S1 = Y(I ) + Y(I2)
                    S3 = Y(I ) - Y(I2)
                    R2 = X(I1) + X(I3)
                    R4 = X(I1) - X(I3)
                    S2 = Y(I1) + Y(I3)
                    S4 = Y(I1) - Y(I3)
                    X(I) = R1 + R2
                    R2   = R1 - R2
                    R1   = R3 - S4
                    R3   = R3 + S4
                    Y(I) = S1 + S2
                    S2   = S1 - S2
                    S1   = S3 + R4
                    S3   = S3 - R4
                    X(I1) = CO1*R3 + SI1*S3
                    Y(I1) = CO1*S3 - SI1*R3
                    X(I2) = CO2*R2 + SI2*S2
                    Y(I2) = CO2*S2 - SI2*R2
                    X(I3) = CO3*R1 + SI3*S1
                    Y(I3) = CO3*S1 - SI3*R1
  30            CONTINUE
                GOTO 20
C-----------------SPECIAL BUTTERFLY FOR  W = J-----------
  50            DO 40 I = J, N, N1
                    I1 = I  + N2
                    I2 = I1 + N2
                    I3 = I2 + N2
                    R1 = X(I ) + X(I2)
                    R3 = X(I ) - X(I2)
                    S1 = Y(I ) + Y(I2)
                    S3 = Y(I ) - Y(I2)
                    R2 = X(I1) + X(I3)
                    R4 = X(I1) - X(I3)
                    S2 = Y(I1) + Y(I3)
                    S4 = Y(I1) - Y(I3)
                    X(I) = R1 + R2
                    Y(I2)=-R1 + R2
                    R1   = R3 - S4
                    R3   = R3 + S4
                    Y(I) = S1 + S2
                    X(I2)= S1 - S2
                    S1   = S3 + R4
                    S3   = S3 - R4
                    X(I1) = (S3 + R3)*C21
                    Y(I1) = (S3 - R3)*C21
                    X(I3) = (S1 - R1)*C21
                    Y(I3) =-(S1 + R1)*C21
  40            CONTINUE
  20        CONTINUE
  10    CONTINUE
C----------DIGIT REVERSE COUNTER----------
C            SAME AS PROGRAM 11
        RETURN
        END
```

```
C       A COOLEY-TUKEY RADIX-4 DIF  FFT PROGRAM
C       THREE BF, MULT BY 1, J,ETC. ARE REMOVED
C       COMPLEX INPUT DATA IN ARRAYS X AND Y
C       LENGTH IS  N = 4 ** M
C       TABLE LOOKUP OF  W  VALUES
C       THE NUMBER OF TEMP VARIABLES IS MINIMUM
C
C         C. S. BURRUS, RICE UNIVERSITY, SEPT 1983
C
C-----------------------------------------------------------
C
        SUBROUTINE  FFT4 (X,Y,N,M,WR,WI)
        REAL X(1), Y(1), WR(1), WI(1)
        DATA  C21 / 0.707106778 /
C
C-------------MAIN FFT LOOPS-----------------------------
C
        N2 = N
        DO 10 K = 1, M
            N1 = N2
            N2 = N2/4
            JT = N2/2 + 1
            DO 1 I = 1, N, N1
                I1 = I  + N2
                I2 = I1 + N2
                I3 = I2 + N2
                R1 = X(I ) + X(I2)
                R2 = X(I ) - X(I2)
                R3 = X(I1) + X(I3)
                X(I ) = R1 + R3
                X(I2) = R1 - R3
                R1 = Y(I ) + Y(I2)
                R4 = Y(I ) - Y(I2)
                R3 = Y(I1) + Y(I3)
                Y(I ) = R1 + R3
                Y(I2) = R1 - R3
                R1 = X(I1) - X(I3)
                R3 = Y(I1) - Y(I3)
                X(I1) = R2 + R3
                X(I3) = R2 - R3
                Y(I1) = R4 - R1
                Y(I3) = R4 + R1
   1        CONTINUE
            IF (K.EQ.M) GOTO 10
            IE = N/N1
            IA1 = 1
            DO 20 J = 2, N2
                IA1 = IA1 + IE
                IF (J.EQ.JT) GOTO 50
                IA2 = IA1 + IA1 - 1
                IA3 = IA2 + IA1 - 1
                CO1  = WR(IA1)
                CO2  = WR(IA2)
                CO3  = WR(IA3)
                SI1  = WI(IA1)
                SI2  = WI(IA2)
                SI3  = WI(IA3)
```

```
                DO 30 I = J, N, N1
                     I1 = I  + N2
                     I2 = I1 + N2
                     I3 = I2 + N2
                     R1 = X(I ) + X(I2)
                     R2 = X(I ) - X(I2)
                     T  = X(I1) + X(I3)
                     X(I)= R1 + T
                     R1  = R1 - T
                     S1 = Y(I ) + Y(I2)
                     S2 = Y(I ) - Y(I2)
                     T  = Y(I1) + Y(I3)
                     Y(I)= S1 + T
                     S1  = S1 - T
                     X(I2) = R1*CO2 + S1*SI2
                     Y(I2) = S1*CO2 - R1*SI2
                     T = Y(I1) - Y(I3)
                     R1 = R2 + T
                     R2 = R2 - T
                     T = X(I1) - X(I3)
                     S1 = S2 - T
                     S2 = S2 + T
                     X(I1) = R1*CO1 + S1*SI1
                     Y(I1) = S1*CO1 - R1*SI1
                     X(I3) = R2*CO3 + S2*SI3
                     Y(I3) = S2*CO3 - R2*SI3
  30            CONTINUE
                GOTO 20
C---------------SPECIAL BUTTERFLY FOR  W = J-------------
  50            DO 40  I = J, N, N1
                     I1 = I  + N2
                     I2 = I1 + N2
                     I3 = I2 + N2
                     R1 = X(I ) + X(I2)
                     R2 = X(I ) - X(I2)
                     S1 = Y(I ) + Y(I2)
                     S2 = Y(I ) - Y(I2)
                     T  = X(I1) + X(I3)
                     X(I ) = T  + R1
                     Y(I2) = T  - R1
                     T  = Y(I1) + Y(I3)
                     Y(I ) = S1 + T
                     X(I2) = S1 - T
                     R1 = X(I1) - X(I3)
                     S1 = Y(I1) - Y(I3)
                     T  = R2 + S1
                     R2 = R2 - S1
                     S1 = S2 - R1
                     S2 = S2 + R1
                     X(I1) = (T  + S1)*C21
                     Y(I1) = (S1 - T )*C21
                     X(I3) = (S2 - R2)*C21
                     Y(I3) =-(S2 + R2)*C21
  40            CONTINUE
  20        CONTINUE
  10    CONTINUE
C-----------DIGIT REVERSE COUNTER----------
C             SAME AS PROGRAM 11
        RETURN
        END
```

```
C        A COOLEY-TUKEY RADIX-4 DIF  FFT PROGRAM
C        THREE BF, MULT BY 1, J,ETC. ARE REMOVED
C        COMPLEX INPUT DATA IN ARRAYS X AND Y
C        LENGTH IS  N = 4 ** M
C        TABLE LOOKUP OF  W  VALUES
C        THE NUMBER OF TEMP VARIABLES IS MINIMUM
C        THREE REAL MULTS PER COMPLEX MULT
C
C          C. S. BURRUS, RICE UNIVERSITY, SEPT 1983
C
C-----------------------------------------------------------------
C
        SUBROUTINE  FFT4 (X,Y,N,M,WR,WI)
        REAL X(1), Y(1), WR(1), WI(1)
        DATA  C21 / 0.707106778 /
C
C--------------MAIN FFT LOOPS-----------------------------
C
        N2 = N
        DO 10 K = 1, M
            N1 = N2
            N2 = N2/4
            JT = N2/2 + 1
            DO 1 I = 1, N, N1
                I1 = I  + N2
                I2 = I1 + N2
                I3 = I2 + N2
                R1 = X(I ) + X(I2)
                R2 = X(I ) - X(I2)
                R3 = X(I1) + X(I3)
                X(I ) = R1 + R3
                X(I2) = R1 - R3
                R1 = Y(I ) + Y(I2)
                R4 = Y(I ) - Y(I2)
                R3 = Y(I1) + Y(I3)
                Y(I ) = R1 + R3
                Y(I2) = R1 - R3
                R1 = X(I1) - X(I3)
                R3 = Y(I1) - Y(I3)
                X(I1) = R2 + R3
                X(I3) = R2 - R3
                Y(I1) = R4 - R1
                Y(I3) = R4 + R1
   1        CONTINUE
            IF (K.EQ.M) GOTO 10
            IE = N/N1
            IA1 = 1
            DO 20 J = 2, N2
                IA1 = IA1 + IE
                IF (J.EQ.JT) GOTO 50
                IA2 = IA1 + IA1 - 1
                IA3 = IA2 + IA1 - 1
                CO1  = WR(IA1)
                CO2  = WR(IA2)
                CO3  = WR(IA3)
                SI1  =-WI(IA1)-CO1
                SI2  =-WI(IA2)-CO2
                SI3  =-WI(IA3)-CO3
                CS1  = WI(IA1)-CO1
                CS2  = WI(IA2)-CO2
                CS3  = WI(IA3)-CO3
                DO 30 I = J, N, N1
```

```
                    I1 = I  + N2
                    I2 = I1 + N2
                    I3 = I2 + N2
                    R1 = X(I ) + X(I2)
                    R2 = X(I ) - X(I2)
                    T  = X(I1) + X(I3)
                    X(I)= R1 + T
                    R1  = R1 - T
                    S1 = Y(I ) + Y(I2)
                    S2 = Y(I ) - Y(I2)
                    T  = Y(I1) + Y(I3)
                    Y(I)= S1 + T
                    S1  = S1 - T
                    T  = (R1 + S1)*CO2
                    Y(I2) = T + R1*SI2
                    X(I2) = T + S1*CS2
                    T = Y(I1) - Y(I3)
                    R1 = R2 + T
                    R2 = R2 - T
                    T = X(I1) - X(I3)
                    S1 = S2 - T
                    S2 = S2 + T
                    T = (R1 + S1)*CO1
                    Y(I1) = T + R1*SI1
                    X(I1) = T + S1*CS1
                    T = (R2 + S2)*CO3
                    Y(I3) = T + R2*SI3
                    X(I3) = T + S2*CS3
 30             CONTINUE
                GOTO 20
C---------------SPECIAL BUTTERFLY FOR  W = J-------------
 50             DO 40  I = J, N, N1
                    I1 = I  + N2
                    I2 = I1 + N2
                    I3 = I2 + N2
                    R1 = X(I ) + X(I2)
                    R2 = X(I ) - X(I2)
                    S1 = Y(I ) + Y(I2)
                    S2 = Y(I ) - Y(I2)
                    T  = X(I1) + X(I3)
                    X(I ) = T  + R1
                    Y(I2) = T  - R1
                    T  = Y(I1) + Y(I3)
                    Y(I ) = S1 + T
                    X(I2) = S1 - T
                    R1 = X(I1) - X(I3)
                    S1 = Y(I1) - Y(I3)
                    T  = R2 + S1
                    R2 = R2 - S1
                    S1 = S2 - R1
                    S2 = S2 + R1
                    X(I1) = (T  + S1)*C21
                    Y(I1) = (S1 - T )*C21
                    X(I3) = (S2 - R2)*C21
                    Y(I3) =-(S2 + R2)*C21
 40             CONTINUE
 20         CONTINUE
 10     CONTINUE
C-----------DIGIT REVERSE COUNTER----------
C            SAME AS PROGRAM 11
        RETURN
        END
```

```
C       A COOLEY-TUKEY RADIX-8 DIF  FFT PROGRAM
C       COMPLEX INPUT DATA IN ARRAYS X AND Y
C       LENGTH IS  N = 8 ** M
C
C         C. S. BURRUS, RICE UNIVERSITY, SEPT 1983
C
C---------------------------------------------------------------
C
        SUBROUTINE  FFT (X,Y,N,M)
        REAL X(1), Y(1)
        C81 = 0.707106778
C
C--------------MAIN FFT LOOPS-----------------------------------
C
        N2 = N
        DO 10 K = 1, M
            N1 = N2
            N2 = N2/8
            E1 = 6.283185307179586/N1
            A = 0
C-------------------MAIN BUTTERFLIES-------------------
            DO 20 J=1, N2
                B     = A + A
                C     = A + B
                D     = A + C
                E     = A + D
                F     = A + E
                G     = A + F
                CO2   = COS(A)
                CO3   = COS(B)
                CO4   = COS(C)
                CO5   = COS(D)
                CO6   = COS(E)
                CO7   = COS(F)
                CO8   = COS(G)
                SI2   = SIN(A)
                SI3   = SIN(B)
                SI4   = SIN(C)
                SI5   = SIN(D)
                SI6   = SIN(E)
                SI7   = SIN(F)
                SI8   = SIN(G)
                A     = J*E1
C---------------BUTTERFLIES WITH SAME W---------------
                DO 30 I1 = J, N, N1
                    I2 = I1 + N2
                    I3 = I2 + N2
                    I4 = I3 + N2
                    I5 = I4 + N2
                    I6 = I5 + N2
                    I7 = I6 + N2
                    I8 = I7 + N2
                    R1 = X(I1) + X(I5)
                    R5 = X(I1) - X(I5)
                    R2 = X(I2) + X(I6)
                    R6 = X(I2) - X(I6)
                    R3 = X(I3) + X(I7)
                    R7 = X(I3) - X(I7)
                    R4 = X(I4) + X(I8)
                    R8 = X(I4) - X(I8)
```

```
                  T1 = R1 - R3
                  R1 = R1 + R3
                  R3 = R2 - R4
                  R2 = R2 + R4
                  X(I1) = R1 + R2
                  R2    = R1 - R2
                  S1 = Y(I1) + Y(I5)
                  S5 = Y(I1) - Y(I5)
                  S2 = Y(I2) + Y(I6)
                  S6 = Y(I2) - Y(I6)
                  S3 = Y(I3) + Y(I7)
                  S7 = Y(I3) - Y(I7)
                  S4 = Y(I4) + Y(I8)
                  S8 = Y(I4) - Y(I8)
                  T2 = S1 - S3
                  S1 = S1 + S3
                  S3 = S2 - S4
                  S2 = S2 + S4
                  Y(I1) = S1 + S2
                  S2    = S1 - S2
                  R1 = T1 + S3
                  T1 = T1 - S3
                  S1 = T2 - R3
                  T2 = T2 + R3
                  X(I5) = CO5*R2 + SI5*S2
                  Y(I5) = CO5*S2 - SI5*R2
                  X(I3) = CO3*R1 + SI3*S1
                  Y(I3) = CO3*S1 - SI3*R1
                  X(I7) = CO7*T1 + SI7*T2
                  Y(I7) = CO7*T2 - SI7*T1
                  R1 = (R6 - R8)*C81
                  R6 = (R6 + R8)*C81
                  S1 = (S6 - S8)*C81
                  S6 = (S6 + S8)*C81
                  T1 = R5 - R1
                  R5 = R5 + R1
                  R8 = R7 - R6
                  R7 = R7 + R6
                  T2 = S5 - S1
                  S5 = S5 + S1
                  S8 = S7 - S6
                  S7 = S7 + S6
                  R1 = R5 + S7
                  R5 = R5 - S7
                  R6 = T1 + S8
                  T1 = T1 - S8
                  S1 = S5 - R7
                  S5 = S5 + R7
                  S6 = T2 - R8
                  T2 = T2 + R8
                  X(I2) = CO2*R1 + SI2*S1
                  Y(I2) = CO2*S1 - SI2*R1
                  X(I8) = CO8*R5 + SI8*S5
                  Y(I8) = CO8*S5 - SI8*R5
                  X(I6) = CO6*R6 + SI6*S6
                  Y(I6) = CO6*S6 - SI6*R6
                  X(I4) = CO4*T1 + SI4*T2
                  Y(I4) = CO4*T2 - SI4*T1
30             CONTINUE
20          CONTINUE
10       CONTINUE
```

```
C-----------DIGIT REVERSE COUNTER----------
  100   J = 1
        N1 = N - 1
        DO 104 I = 1, N1
            IF (I.GE.J) GOTO 101
            T1 = X(J)
            X(J) = X(I)
            X(I) = T1
            T1 = Y(J)
            Y(J) = Y(I)
            Y(I) = T1
 101        K = N/8
 102        IF (K*7.GE.J) GOTO 103
                J = J - K*7
                K = K/8
                GOTO 102
 103        J = J + K
 104    CONTINUE
        RETURN
        END
```

```
C       A COOLEY-TUKEY RADIX-8 DIF  FFT PROGRAM
C       TWO BF, TABLE LOOKUP OF  W  VALUES
C       COMPLEX INPUT DATA IN ARRAYS X AND Y
C       TWIDDLE FACTORS ARE STORED IN TABLE
C       LENGTH IS  N = 8 ** M
C
C         C. S. BURRUS, RICE UNIVERSITY, SEPT 1983
C
C---------------------------------------------------------------
C
        SUBROUTINE  FFT8 (X,Y,N,M,WR,WI)
        REAL X(1), Y(1), WR(1), WI(1)
        C81 = 0.707106778
C-------------MAIN FFT LOOPS-----------------------------------
        N2 = N
        DO 10 K = 1, M
            N1 = N2
            N2 = N2/8
            DO 1 I1 = 1, N, N1
                    I2 = I1 + N2
                    I3 = I2 + N2
                    I4 = I3 + N2
                    I5 = I4 + N2
                    I6 = I5 + N2
                    I7 = I6 + N2
                    I8 = I7 + N2
                    R1 = X(I1) + X(I5)
                    R5 = X(I1) - X(I5)
                    R2 = X(I2) + X(I6)
                    R6 = X(I2) - X(I6)
                    R3 = X(I3) + X(I7)
                    R7 = X(I3) - X(I7)
                    R4 = X(I4) + X(I8)
                    R8 = X(I4) - X(I8)
                    T1 = R1 - R3
                    R1 = R1 + R3
                    R3 = R2 - R4
                    R2 = R2 + R4
                    X(I1) = R1 + R2
                    X(I5) = R1 - R2
                    R1 = Y(I1) + Y(I5)
                    S5 = Y(I1) - Y(I5)
                    R2 = Y(I2) + Y(I6)
                    S6 = Y(I2) - Y(I6)
                    S3 = Y(I3) + Y(I7)
                    S7 = Y(I3) - Y(I7)
                    R4 = Y(I4) + Y(I8)
                    S8 = Y(I4) - Y(I8)
                    T2 = R1 - S3
                    R1 = R1 + S3
                    S3 = R2 - R4
                    R2 = R2 + R4
                    Y(I1) = R1 + R2
                    Y(I5) = R1 - R2
                    X(I3) = T1 + S3
                    X(I7) = T1 - S3
                    Y(I3) = T2 - R3
                    Y(I7) = T2 + R3
```

```
                  R1 = (R6 - R8)*C81
                  R6 = (R6 + R8)*C81
                  R2 = (S6 - S8)*C81
                  S6 = (S6 + S8)*C81
                  T1 = R5 - R1
                  R5 = R5 + R1
                  R8 = R7 - R6
                  R7 = R7 + R6
                  T2 = S5 - R2
                  S5 = S5 + R2
                  S8 = S7 - S6
                  S7 = S7 + S6
                  X(I2) = R5 + S7
                  X(I8) = R5 - S7
                  X(I6) = T1 + S8
                  X(I4) = T1 - S8
                  Y(I2) = S5 - R7
                  Y(I8) = S5 + R7
                  Y(I6) = T2 - R8
                  Y(I4) = T2 + R8
 1          CONTINUE
            IF (K.EQ.M) GOTO 10
            IE = N/N1 - 1
            IA1 = 1
            DO 20 J = 2, N2
               ID   = IA1 + IE
               IA1  = 1   + ID
               IA2  = IA1 + ID
               IA3  = IA2 + ID
               IA4  = IA3 + ID
               IA5  = IA4 + ID
               IA6  = IA5 + ID
               IA7  = IA6 + ID
               CO2  = WR(IA1)
               CO3  = WR(IA2)
               CO4  = WR(IA3)
               CO5  = WR(IA4)
               CO6  = WR(IA5)
               CO7  = WR(IA6)
               CO8  = WR(IA7)
               SI2  = WI(IA1)
               SI3  = WI(IA2)
               SI4  = WI(IA3)
               SI5  = WI(IA4)
               SI6  = WI(IA5)
               SI7  = WI(IA6)
               SI8  = WI(IA7)
C---------------BUTTERFLIES WITH SAME W---------------
               DO 30 I1 = J, N, N1
                  I2 = I1 + N2
                  I3 = I2 + N2
                  I4 = I3 + N2
                  I5 = I4 + N2
                  I6 = I5 + N2
                  I7 = I6 + N2
                  I8 = I7 + N2
                  R1 = X(I1) + X(I5)
                  R5 = X(I1) - X(I5)
                  R2 = X(I2) + X(I6)
                  R6 = X(I2) - X(I6)
                  R3 = X(I3) + X(I7)
                  R7 = X(I3) - X(I7)
                  R4 = X(I4) + X(I8)
                  R8 = X(I4) - X(I8)
```

```
                  T1 = R1 - R3
                  R1 = R1 + R3
                  R3 = R2 - R4
                  R2 = R2 + R4
                  X(I1) = R1 + R2
                  R2    = R1 - R2
                  S1 = Y(I1) + Y(I5)
                  S5 = Y(I1) - Y(I5)
                  S2 = Y(I2) + Y(I6)
                  S6 = Y(I2) - Y(I6)
                  S3 = Y(I3) + Y(I7)
                  S7 = Y(I3) - Y(I7)
                  S4 = Y(I4) + Y(I8)
                  S8 = Y(I4) - Y(I8)
                  T2 = S1 - S3
                  S1 = S1 + S3
                  S3 = S2 - S4
                  S2 = S2 + S4
                  Y(I1) = S1 + S2
                  S2    = S1 - S2
                  R1 = T1 + S3
                  T1 = T1 - S3
                  S1 = T2 - R3
                  T2 = T2 + R3
                  X(I5) = CO5*R2 + SI5*S2
                  Y(I5) = CO5*S2 - SI5*R2
                  X(I3) = CO3*R1 + SI3*S1
                  Y(I3) = CO3*S1 - SI3*R1
                  X(I7) = CO7*T1 + SI7*T2
                  Y(I7) = CO7*T2 - SI7*T1
                  R1 = (R6 - R8)*C81
                  R6 = (R6 + R8)*C81
                  S1 = (S6 - S8)*C81
                  S6 = (S6 + S8)*C81
                  T1 = R5 - R1
                  R5 = R5 + R1
                  R8 = R7 - R6
                  R7 = R7 + R6
                  T2 = S5 - S1
                  S5 = S5 + S1
                  S8 = S7 - S6
                  S7 = S7 + S6
                  R1 = R5 + S7
                  R5 = R5 - S7
                  R6 = T1 + S8
                  T1 = T1 - S8
                  S1 = S5 - R7
                  S5 = S5 + R7
                  S6 = T2 - R8
                  T2 = T2 + R8
                  X(I2) = CO2*R1 + SI2*S1
                  Y(I2) = CO2*S1 - SI2*R1
                  X(I8) = CO8*R5 + SI8*S5
                  Y(I8) = CO8*S5 - SI8*R5
                  X(I6) = CO6*R6 + SI6*S6
                  Y(I6) = CO6*S6 - SI6*R6
                  X(I4) = CO4*T1 + SI4*T2
                  Y(I4) = CO4*T2 - SI4*T1
   30          CONTINUE
   20       CONTINUE
   10    CONTINUE
C-----------DIGIT REVERSE COUNTER----------
C           SAME AS PROGRAM 15
         RETURN
         END
```

```
C    A PRIME FACTOR FFT PROGRAM
C    COMPLEX INPUT DATA IN ARRAYS  X AND Y
C    LENGTH  N  WITH  M  FACTORS IN ARRAY  NI
C        N = NI(1)*NI(2)*...*NI(M)
C
C    PROGRAM BY  C.S. BURRUS,  RICE UNIVERSITY
C
C                    SEPT 1983
C
C------------------------------------------------------------
C
        SUBROUTINE PFA(X,Y,A,B,N,M,NI,UNSC)
        INTEGER  NI(4), I(16), UNSC
        REAL X(1), Y(1), A(1), B(1)
C
        DATA  C31, C32  / -0.86602540,-1.50000000 /
        DATA  C51, C52  /  0.95105652,-1.53884180 /
        DATA  C53, C54  / -0.36327126, 0.55901699 /
        DATA  C55       / -1.25  /
        DATA  C71, C72  / -1.16666667,-0.79015647 /
        DATA  C73, C74  /  0.055854267, 0.7343022 /
        DATA  C75, C76  /  0.44095855,-0.34087293 /
        DATA  C77, C78  /  0.53396936, 0.87484229 /
        DATA  C81       /  0.70710678 /
        DATA  C95       / -0.50000000 /
        DATA  C92, C93  /  0.93969262, -0.17364818 /
        DATA  C94, C96  /  0.76604444, -0.34202014 /
        DATA  C97, C98  / -0.98480775, -0.64278761 /
        DATA  C162,C163 /  0.38268343,  1.30656297 /
        DATA  C164,C165 /  0.54119610,  0.92387953 /
C
C-----------------NESTED LOOPS-----------------------------
C
        DO 10 K=1, M
           N1 = NI(K)
           N2 = N/N1
           DO 20 J=1, N, N1
              IT   = J
              I(1) = J
              DO 30 L=2, N1
                 IT = IT + N2
                 IF (IT.GT.N)  IT = IT - N
                 I(L) = IT
   30         CONTINUE
              GOTO (20,102,103,104,105,20,107,108,109,
     +                 20,20,20,20,20,20,116),N1
   20      CONTINUE
   10   CONTINUE
C
C----------------UNSCRAMBLE---------------------------------
        L = 1
        DO 2 K = 1, N
           A(K) = X(L)
           B(K) = Y(L)
           L = L + UNSC
           IF (L.GT.N) L = L - N
    2   CONTINUE
C
        RETURN
```

```
C
C---------------WFTA N=2----------------------------------
C
  102    R1         = X(I(1))
         X(I(1))    = R1 + X(I(2))
         X(I(2))    = R1 - X(I(2))
C
         R1         = Y(I(1))
         Y(I(1)) = R1 + Y(I(2))
         Y(I(2)) = R1 - Y(I(2))
C
         GOTO 20
C
C---------------WFTA N=3----------------------------------
C
  103    R2 = (X(I(2)) - X(I(3))) * C31
         R1 =  X(I(2)) + X(I(3))
         X(I(1))= X(I(1)) + R1
         R1      = X(I(1)) + R1 * C32
C
         S2 = (Y(I(2)) - Y(I(3))) * C31
         S1 =  Y(I(2)) + Y(I(3))
         Y(I(1))= Y(I(1)) + S1
         S1      = Y(I(1)) + S1 * C32
C
         X(I(2)) = R1 - S2
         X(I(3)) = R1 + S2
         Y(I(2)) = S1 + R2
         Y(I(3)) = S1 - R2
C
         GOTO 20
C
C---------------WFTA N=4----------------------------------
C
  104    R1 = X(I(1)) + X(I(3))
         T1 = X(I(1)) - X(I(3))
         R2 = X(I(2)) + X(I(4))
         X(I(1)) = R1 + R2
         X(I(3)) = R1 - R2
C
         R1 = Y(I(1)) + Y(I(3))
         T2 = Y(I(1)) - Y(I(3))
         R2 = Y(I(2)) + Y(I(4))
         Y(I(1)) = R1 + R2
         Y(I(3)) = R1 - R2
C
         R1 = X(I(2)) - X(I(4))
         R2 = Y(I(2)) - Y(I(4))
C
         X(I(2)) = T1 + R2
         X(I(4)) = T1 - R2
         Y(I(2)) = T2 - R1
         Y(I(4)) = T2 + R1
C
         GOTO 20
C
```

```
C
C----------------WFTA N=5----------------------------------
C
  105   R1 = X(I(2)) + X(I(5))
        R4 = X(I(2)) - X(I(5))
        R3 = X(I(3)) + X(I(4))
        R2 = X(I(3)) - X(I(4))
C
        T = (R1 - R3) * C54
        R1 = R1 + R3
        X(I(1)) = X(I(1)) + R1
        R1      = X(I(1)) + R1 * C55
C
        R3 = R1 - T
        R1 = R1 + T
C
        T = (R4 + R2) * C51
        R4 =  T + R4 * C52
        R2 =  T + R2 * C53
C
        S1 = Y(I(2)) + Y(I(5))
        S4 = Y(I(2)) - Y(I(5))
        S3 = Y(I(3)) + Y(I(4))
        S2 = Y(I(3)) - Y(I(4))
C
        T = (S1 - S3) * C54
        S1 = S1 + S3
        Y(I(1)) = Y(I(1)) + S1
        S1      = Y(I(1)) + S1 * C55
C
        S3 =  S1 - T
        S1 =  S1 + T
C
        T = (S4 + S2) * C51
        S4 =  T + S4 * C52
        S2 =  T + S2 * C53
C
        X(I(2)) = R1 + S2
        X(I(5)) = R1 - S2
        X(I(3)) = R3 - S4
        X(I(4)) = R3 + S4
C
        Y(I(2)) = S1 - R2
        Y(I(5)) = S1 + R2
        Y(I(3)) = S3 + R4
        Y(I(4)) = S3 - R4
C
        GOTO 20
C
```

```
C-----------------WFTA N=7----------------------------------
C
  107   R1 = X(I(2)) + X(I(7))
        R6 = X(I(2)) - X(I(7))
        S1 = Y(I(2)) + Y(I(7))
        S6 = Y(I(2)) - Y(I(7))
        R2 = X(I(3)) + X(I(6))
        R5 = X(I(3)) - X(I(6))
        S2 = Y(I(3)) + Y(I(6))
        S5 = Y(I(3)) - Y(I(6))
        R3 = X(I(4)) + X(I(5))
        R4 = X(I(4)) - X(I(5))
        S3 = Y(I(4)) + Y(I(5))
        S4 = Y(I(4)) - Y(I(5))
C
        T3 = (R1 - R2) * C74
        T  = (R1 - R3) * C72
        R1 = R1 + R2 + R3
        X(I(1)) = X(I(1)) + R1
        R1      = X(I(1)) + R1 * C71
        R2 =(R3 - R2) * C73
        R3 = R1 - T + R2
        R2 = R1 - R2 - T3
        R1 = R1 + T + T3
        T = (R6 - R5) * C78
        T3 =(R6 + R4) * C76
        R6 =(R6 + R5 - R4) * C75
        R5 =(R5 + R4) * C77
        R4 = R6 - T3 + R5
        R5 = R6 - R5 - T
        R6 = R6 + T3 + T
C
        T3 = (S1 - S2) * C74
        T  = (S1 - S3) * C72
        S1 =  S1 + S2 + S3
        Y(I(1)) = Y(I(1)) + S1
        S1      = Y(I(1)) + S1 * C71
        S2 =(S3 - S2) * C73
        S3 = S1 - T  + S2
        S2 = S1 - S2 - T3
        S1 = S1 + T  + T3
        T  = (S6 - S5) * C78
        T3 = (S6 + S4) * C76
        S6 = (S6 + S5 - S4) * C75
        S5 = (S5 + S4) * C77
        S4 = S6 - T3 + S5
        S5 = S6 - S5 - T
        S6 = S6 + T3 + T
C
        X(I(2)) = R3 + S4
        X(I(7)) = R3 - S4
        X(I(3)) = R1 + S6
        X(I(6)) = R1 - S6
        X(I(4)) = R2 - S5
        X(I(5)) = R2 + S5
        Y(I(4)) = S2 + R5
        Y(I(5)) = S2 - R5
        Y(I(2)) = S3 - R4
        Y(I(7)) = S3 + R4
        Y(I(3)) = S1 - R6
        Y(I(6)) = S1 + R6
C
        GOTO 20
```

```
C
C-----------------WFTA N=8---------------------------------
C
  108   R1 = X(I(1)) + X(I(5))
        R2 = X(I(1)) - X(I(5))
        R3 = X(I(2)) + X(I(8))
        R4 = X(I(2)) - X(I(8))
        R5 = X(I(3)) + X(I(7))
        R6 = X(I(3)) - X(I(7))
        R7 = X(I(4)) + X(I(6))
        R8 = X(I(4)) - X(I(6))
        T1 = R1 + R5
        T2 = R1 - R5
        T3 = R3 + R7
        R3 =(R3 - R7) * C81
        X(I(1)) = T1 + T3
        X(I(5)) = T1 - T3
        T1 = R2 + R3
        T3 = R2 - R3
        S1 = R4 - R8
        R4 =(R4 + R8) * C81
        S2 = R4 + R6
        S3 = R4 - R6
        R1 = Y(I(1)) + Y(I(5))
        R2 = Y(I(1)) - Y(I(5))
        R3 = Y(I(2)) + Y(I(8))
        R4 = Y(I(2)) - Y(I(8))
        R5 = Y(I(3)) + Y(I(7))
        R6 = Y(I(3)) - Y(I(7))
        R7 = Y(I(4)) + Y(I(6))
        R8 = Y(I(4)) - Y(I(6))
        T4 = R1 + R5
        R1 = R1 - R5
        R5 = R3 + R7
        R3 =(R3 - R7) * C81
        Y(I(1)) = T4 + R5
        Y(I(5)) = T4 - R5
        R5 = R2 + R3
        R2 = R2 - R3
        R3 = R4 - R8
        R4 =(R4 + R8) * C81
        R7 = R4 + R6
        R4 = R4 - R6
        X(I(2)) = T1 + R7
        X(I(8)) = T1 - R7
        X(I(3)) = T2 + R3
        X(I(7)) = T2 - R3
        X(I(4)) = T3 + R4
        X(I(6)) = T3 - R4
        Y(I(2)) = R5 - S2
        Y(I(8)) = R5 + S2
        Y(I(3)) = R1 - S1
        Y(I(7)) = R1 + S1
        Y(I(4)) = R2 - S3
        Y(I(6)) = R2 + S3
C
        GOTO 20
C
```

```
C
C-----------------WFTA N=9---------------------------------
C
  109   R1 = X(I(2)) + X(I(9))
        R2 = X(I(2)) - X(I(9))
        R3 = X(I(3)) + X(I(8))
        R4 = X(I(3)) - X(I(8))
        R5 = X(I(4)) + X(I(7))
        T8 =(X(I(4)) - X(I(7))) * C31
        R7 = X(I(5)) + X(I(6))
        R8 = X(I(5)) - X(I(6))
        T0 = X(I(1)) + R5
        T7 = X(I(1)) + R5 * C95
        R5 = R1 + R3 + R7
        X(I(1)) = T0 + R5
        T5 = T0 + R5 * C95
        T3 = (R3 - R7) * C92
        R7 = (R1 - R7) * C93
        R3 = (R1 - R3) * C94
        T1 = T7 + T3 + R3
        T3 = T7 - T3 - R7
        T7 = T7 + R7 - R3
        T6 = (R2 - R4 + R8) * C31
        T4 = (R4 + R8) * C96
        R8 = (R2 - R8) * C97
        R2 = (R2 + R4) * C98
        T2 = T8 + T4 + R2
        T4 = T8 - T4 - R8
        T8 = T8 + R8 - R2
C
        R1 = Y(I(2)) + Y(I(9))
        R2 = Y(I(2)) - Y(I(9))
        R3 = Y(I(3)) + Y(I(8))
        R4 = Y(I(3)) - Y(I(8))
        R5 = Y(I(4)) + Y(I(7))
        R6 =(Y(I(4)) - Y(I(7))) * C31
        R7 = Y(I(5)) + Y(I(6))
        R8 = Y(I(5)) - Y(I(6))
        T0 = Y(I(1)) + R5
        T9 = Y(I(1)) + R5 * C95
        R5 = R1 + R3 + R7
        Y(I(1)) = T0 + R5
        R5 = T0 + R5 * C95
        T0 = (R3 - R7) * C92
        R7 = (R1 - R7) * C93
        R3 = (R1 - R3) * C94
        R1 = T9 + T0 + R3
        T0 = T9 - T0 - R7
        R7 = T9 + R7 - R3
        R9 = (R2 - R4 + R8) * C31
        R3 = (R4 + R8) * C96
        R8 = (R2 - R8) * C97
        R4 = (R2 + R4) * C98
        R2 = R6 + R3 + R4
        R3 = R6 - R8 - R3
        R8 = R6 + R8 - R4
```

```
C
      X(I(2)) = T1 - R2
      X(I(9)) = T1 + R2
      Y(I(2)) = R1 + T2
      Y(I(9)) = R1 - T2
      X(I(3)) = T3 + R3
      X(I(8)) = T3 - R3
      Y(I(3)) = T0 - T4
      Y(I(8)) = T0 + T4
      X(I(4)) = T5 - R9
      X(I(7)) = T5 + R9
      Y(I(4)) = R5 + T6
      Y(I(7)) = R5 - T6
      X(I(5)) = T7 - R8
      X(I(6)) = T7 + R8
      Y(I(5)) = R7 + T8
      Y(I(6)) = R7 - T8
C
      GOTO 20
C

C-----------------WFTA N=16----------------------------------
C
 116  R1 = X(I(1)) + X(I(9))
      R2 = X(I(1)) - X(I(9))
      R3 = X(I(2)) + X(I(10))
      R4 = X(I(2)) - X(I(10))
      R5 = X(I(3)) + X(I(11))
      R6 = X(I(3)) - X(I(11))
      R7 = X(I(4)) + X(I(12))
      R8 = X(I(4)) - X(I(12))
      R9 = X(I(5)) + X(I(13))
      R10= X(I(5)) - X(I(13))
      R11 = X(I(6)) + X(I(14))
      R12 = X(I(6)) - X(I(14))
      R13 = X(I(7)) + X(I(15))
      R14 = X(I(7)) - X(I(15))
      R15 = X(I(8)) + X(I(16))
      R16 = X(I(8)) - X(I(16))
      T1 = R1 + R9
      T2 = R1 - R9
      T3 = R3 + R11
      T4 = R3 - R11
      T5 = R5 + R13
      T6 = R5 - R13
      T7 = R7 + R15
      T8 = R7 - R15
      R1 = T1 + T5
      R3 = T1 - T5
      R5 = T3 + T7
      R7 = T3 - T7
      X(I( 1)) = R1 + R5
      X(I( 9)) = R1 - R5
      T1 = C81 * (T4 + T8)
      T5 = C81 * (T4 - T8)
      R9 = T2 + T5
      R11= T2 - T5
      R13 = T6 + T1
      R15 = T6 - T1
      T1 = R4 + R16
      T2 = R4 - R16
      T3 = C81 * (R6 + R14)
      T4 = C81 * (R6 - R14)
      T5 = R8 + R12
      T6 = R8 - R12
      T7 = C162 * (T2 - T6)
      T2 = C163 * T2 - T7
```

```
T6 = C164 * T6 - T7
T7 = R2 + T4
T8 = R2 - T4
R2 = T7 + T2
R4 = T7 - T2
R6 = T8 + T6
R8 = T8 - T6
T7 = C165 * (T1 + T5)
T2 = T7 - C164 * T1
T4 = T7 - C163 * T5
T6 = R10 + T3
T8 = R10 - T3
R10 = T6 + T2
R12 = T6 - T2
R14 = T8 + T4
R16 = T8 - T4
R1 = Y(I(1)) + Y(I(9))
S2 = Y(I(1)) - Y(I(9))
S3 = Y(I(2)) + Y(I(10))
S4 = Y(I(2)) - Y(I(10))
R5 = Y(I(3)) + Y(I(11))
S6 = Y(I(3)) - Y(I(11))
S7 = Y(I(4)) + Y(I(12))
S8 = Y(I(4)) - Y(I(12))
S9 = Y(I(5)) + Y(I(13))
S10= Y(I(5)) - Y(I(13))
S11 = Y(I(6)) + Y(I(14))
S12 = Y(I(6)) - Y(I(14))
S13 = Y(I(7)) + Y(I(15))
S14 = Y(I(7)) - Y(I(15))
S15 = Y(I(8)) + Y(I(16))
S16 = Y(I(8)) - Y(I(16))
T1 = R1 + S9
T2 = R1 - S9
T3 = S3 + S11
T4 = S3 - S11
T5 = R5 + S13
T6 = R5 - S13
T7 = S7 + S15
T8 = S7 - S15
R1 = T1 + T5
S3 = T1 - T5
R5 = T3 + T7
S7 = T3 - T7
Y(I( 1)) = R1 + R5
Y(I( 9)) = R1 - R5
X(I( 5)) = R3 + S7
X(I(13)) = R3 - S7
Y(I( 5)) = S3 - R7
Y(I(13)) = S3 + R7
T1 = C81 * (T4 + T8)
T5 = C81 * (T4 - T8)
S9 = T2 + T5
S11= T2 - T5
S13 = T6 + T1
S15 = T6 - T1
T1 = S4 + S16
T2 = S4 - S16
T3 = C81 * (S6 + S14)
T4 = C81 * (S6 - S14)
T5 = S8 + S12
T6 = S8 - S12
T7 = C162 * (T2 - T6)
T2 = C163 * T2 - T7
T6 = C164 * T6 - T7
T7 = S2 + T4
T8 = S2 - T4
S2 = T7 + T2
```

```
      S4 = T7 - T2
      S6 = T8 + T6
      S8 = T8 - T6
      T7 = C165 * (T1 + T5)
      T2 = T7 - C164 * T1
      T4 = T7 - C163 * T5
      T6 = S10 + T3
      T8 = S10 - T3
      S10 = T6 + T2
      S12 = T6 - T2
      S14 = T8 + T4
      S16 = T8 - T4
      X(I( 2)) = R2 + S10
      X(I(16)) = R2 - S10
      Y(I( 2)) = S2 - R10
      Y(I(16)) = S2 + R10
      X(I( 3)) = R9 + S13
      X(I(15)) = R9 - S13
      Y(I( 3)) = S9 - R13
      Y(I(15)) = S9 + R13
      X(I( 4)) = R8 - S16
      X(I(14)) = R8 + S16
      Y(I( 4)) = S8 + R16
      Y(I(14)) = S8 - R16
      X(I( 6)) = R6 + S14
      X(I(12)) = R6 - S14
      Y(I( 6)) = S6 - R14
      Y(I(12)) = S6 + R14
      X(I( 7)) = R11 - S15
      X(I(11)) = R11 + S15
      Y(I( 7)) = S11 + R15
      Y(I(11)) = S11 - R15
      X(I( 8)) = R4 - S12
      X(I(10)) = R4 + S12
      Y(I( 8)) = S4 + R12
      Y(I(10)) = S4 - R12
C
      GOTO 20
C
      END
```

```
C
C   A PRIME FACTOR FFT PROGRAM
C   IN-PLACE AND IN-ORDER USING SEPARATE OUTPUT POINTER
C   COMPLEX INPUT DATA IN ARRAYS  X AND Y
C   LENGTH  N  WITH  M  FACTORS IN ARRAY  NI
C        N = NI(1)*NI(2)*...*NI(M)
C
C   PROGRAM BY  C. S. BURRUS,  RICE UNIVERSITY
C
C                     SEPT 1983
C
C---------------------------------------------------------
C
      SUBROUTINE PFA(X,Y,N,M,NI)
C
      INTEGER  NI(4), I(16), IP(16), LP(16)
      REAL X(1), Y(1)
C
      DATA  C31, C32  / -0.86602540,-1.50000000 /
      DATA  C51, C52  /  0.95105652,-1.53884180 /
      DATA  C53, C54  / -0.36327126, 0.55901699 /
      DATA  C55       / -1.25  /
C
C----------------NESTED LOOPS----------------------------------
C
      DO 10 K=1, M
         N1 = NI(K)
         N2 = N/N1
C
         L  = 1
         N3 = N2 - N1*(N2/N1)
         DO 15 J = 2, N1
            L = L + N3
            IF (L.GT.N1) L = L - N1
            LP(J) = L
   15    CONTINUE
C
         DO 20 J=1, N, N1
            IT   = J
            I(1) = J
            IP(1)= J
            DO 30 L=2, N1
               IT = IT + N2
               IF (IT.GT.N)  IT = IT - N
               I(L) = IT
               IP(LP(L)) = IT
   30       CONTINUE
            GOTO (20,102,103,104,105), N1
   20    CONTINUE
   10 CONTINUE
      RETURN
C
```

```
C
C----------------WFTA N=2--------------------------------
C
  102   R1         = X(I(1))
        X(I(1))    = R1 + X(I(2))
        X(I(2))    = R1 - X(I(2))
C
        R1         = Y(I(1))
        Y(IP(1))   = R1 + Y(I(2))
        Y(IP(2))   = R1 - Y(I(2))
C
        GOTO 20
C
C----------------WFTA N=3--------------------------------
C
  103   R2 = (X(I(2)) - X(I(3))) * C31
        R1 =  X(I(2)) + X(I(3))
        X(I(1))= X(I(1)) + R1
        R1      = X(I(1)) + R1 * C32
C
        S2 = (Y(I(2)) - Y(I(3))) * C31
        S1 =  Y(I(2)) + Y(I(3))
        Y(I(1))= Y(I(1)) + S1
        S1      = Y(I(1)) + S1 * C32
C
        X(IP(2)) = R1 - S2
        X(IP(3)) = R1 + S2
        Y(IP(2)) = S1 + R2
        Y(IP(3)) = S1 - R2
C
        GOTO 20
C
C----------------WFTA N=4--------------------------------
C
  104   R1 = X(I(1)) + X(I(3))
        T1 = X(I(1)) - X(I(3))
        R2 = X(I(2)) + X(I(4))
        X(IP(1)) = R1 + R2
        X(IP(3)) = R1 - R2
C
        R1 = Y(I(1)) + Y(I(3))
        T2 = Y(I(1)) - Y(I(3))
        R2 = Y(I(2)) + Y(I(4))
        Y(IP(1)) = R1 + R2
        Y(IP(3)) = R1 - R2
C
        R1 = X(I(2)) - X(I(4))
        R2 = Y(I(2)) - Y(I(4))
C
        X(IP(2)) = T1 + R2
        X(IP(4)) = T1 - R2
        Y(IP(2)) = T2 - R1
        Y(IP(4)) = T2 + R1
C
        GOTO 20
C
```

```
C
C----------------WFTA N=5--------------------------------
C
  105   R1 = X(I(2)) + X(I(5))
        R4 = X(I(2)) - X(I(5))
        R3 = X(I(3)) + X(I(4))
        R2 = X(I(3)) - X(I(4))
C
        T = (R1 - R3) * C54
        R1 = R1 + R3
        X(I(1)) = X(I(1)) + R1
        R1      = X(I(1)) + R1 * C55
C
        R3 = R1 - T
        R1 = R1 + T
C
        T = (R4 + R2) * C51
        R4 =  T + R4 * C52
        R2 =  T + R2 * C53
C
        S1 = Y(I(2)) + Y(I(5))
        S4 = Y(I(2)) - Y(I(5))
        S3 = Y(I(3)) + Y(I(4))
        S2 = Y(I(3)) - Y(I(4))
C
        T = (S1 - S3) * C54
        S1 = S1 + S3
        Y(I(1)) = Y(I(1)) + S1
        S1      = Y(I(1)) + S1 * C55
C
        S3 =  S1 - T
        S1 =  S1 + T
C
        T = (S4 + S2) * C51
        S4 =  T + S4 * C52
        S2 =  T + S2 * C53
C
        X(IP(2)) = R1 + S2
        X(IP(5)) = R1 - S2
        X(IP(3)) = R3 - S4
        X(IP(4)) = R3 + S4
C
        Y(IP(2)) = S1 - R2
        Y(IP(5)) = S1 + R2
        Y(IP(3)) = S3 + R4
        Y(IP(4)) = S3 - R4
C
        GOTO 20
        END
```

```
C
C   A PRIME FACTOR FFT PROGRAM
C   IN-PLACE AND IN-ORDER
C   COMPLEX INPUT DATA IN ARRAYS  X AND Y
C   WRITTEN SPECIFICALLY FOR LENGTH  15 = 3*5
C
C   PROGRAM BY  C. S. BURRUS,  RICE UNIVERSITY
C
C                  SEPT 1983
C
C-----------------------------------------------------
C
      SUBROUTINE PFA(X,Y,N,M,NI)
C
      INTEGER  NI(4), I(16)
      REAL X(1), Y(1)
C
      DATA  C31, C32  / -0.86602540,-1.50000000 /
      DATA  C51, C52  /  0.95105652,-1.53884180 /
      DATA  C53, C54  / -0.36327126, 0.55901699 /
      DATA  C55       / -1.25  /
C
C----------------NESTED LOOPS----------------------------------
C
      DO 10 K = 1, M
         N1 = NI(K)
         N2 = N/N1
         DO 20 J = 1, N, N1
            IT   = J
            I(1) = J
            DO 30 L = 2, N1
               IT = IT + N2
               IF (IT.GT.N)  IT = IT - N
               I(L) = IT
   30       CONTINUE
            GOTO (20,20,103,20,105), N1
   20    CONTINUE
   10 CONTINUE
      RETURN
C
C---------------WFTA N=3--------------------------------
C
  103 R2 = (X(I(2)) - X(I(3))) * C31
      R1 =  X(I(2)) + X(I(3))
      X(I(1))= X(I(1)) + R1
      R1     = X(I(1)) + R1 * C32
C
      S2 = (Y(I(2)) - Y(I(3))) * C31
      S1 =  Y(I(2)) + Y(I(3))
      Y(I(1))= Y(I(1)) + S1
      S1     = Y(I(1)) + S1 * C32
C
      X(I(3)) = R1 - S2
      X(I(2)) = R1 + S2
      Y(I(3)) = S1 + R2
      Y(I(2)) = S1 - R2
C
      GOTO 20
C
```

```
C-----------------WFTA N=5----------------------------------
C
  105   R1 = X(I(2)) + X(I(5))
        R4 = X(I(2)) - X(I(5))
        R3 = X(I(3)) + X(I(4))
        R2 = X(I(3)) - X(I(4))
C
        T = (R1 - R3) * C54
        R1 = R1 + R3
        X(I(1)) = X(I(1)) + R1
        R1      = X(I(1)) + R1 * C55
C
        R3 = R1 - T
        R1 = R1 + T
C
        T = (R4 + R2) * C51
        R4 =  T + R4 * C52
        R2 =  T + R2 * C53
C
        S1 = Y(I(2)) + Y(I(5))
        S4 = Y(I(2)) - Y(I(5))
        S3 = Y(I(3)) + Y(I(4))
        S2 = Y(I(3)) - Y(I(4))
C
        T = (S1 - S3) * C54
        S1 = S1 + S3
        Y(I(1)) = Y(I(1)) + S1
        S1      = Y(I(1)) + S1 * C55
C
        S3 =  S1 - T
        S1 =  S1 + T
C
        T = (S4 + S2) * C51
        S4 =  T + S4 * C52
        S2 =  T + S2 * C53
C
        X(I(3)) = R1 + S2
        X(I(4)) = R1 - S2
        X(I(5)) = R3 - S4
        X(I(2)) = R3 + S4
C
        Y(I(3)) = S1 - R2
        Y(I(4)) = S1 + R2
        Y(I(5)) = S3 + R4
        Y(I(2)) = S3 - R4
C
        GOTO 20
        END
```

```
C        A DUHAMEL-HOLLMAN SPLIT-RADIX DIF FFT
C        REF: ELECTRONICS LETTERS, JAN. 5, 1984
C        COMPLEX INPUT AND OUTPUT DATA IN ARRAYS X AND Y
C        LENGTH IS N = 2 ** M, OUTPUT IN BIT-REVERSED ORDER
C        C.S. BURRUS, RICE UNIVERSITY, DEC. 1984
C-----------------------------------------------------------
C
         SUBROUTINE FFT(X,Y,N,M)
         REAL X(1), Y(1)
C
         N2 = 2*N
         DO 10 K = 1, M-1
             N2 = N2/2
             N4 = N2/4
             E  = 6.283185307179586/N2
             A  = 0
             DO 20 J = 1, N4
                 A3  = 3*A
                 CC1 = COS(A)
                 SS1 = SIN(A)
                 CC3 = COS(A3)
                 SS3 = SIN(A3)
                 A   = J*E
                 IS  = J
                 ID  = 2*N2
                 DO  30 I0 = IS, N-1, ID
                     I1 = I0 + N4
                     I2 = I1 + N4
                     I3 = I2 + N4
                     R1     = X(I0) - X(I2)
                     X(I0)  = X(I0) + X(I2)
                     R2     = X(I1) - X(I3)
                     X(I1)  = X(I1) + X(I3)
                     S1     = Y(I0) - Y(I2)
                     Y(I0)  = Y(I0) + Y(I2)
                     S2     = Y(I1) - Y(I3)
                     Y(I1)  = Y(I1) + Y(I3)
                     S3     = R1 - S2
                     R1     = R1 + S2
                     S2     = R2 - S1
                     R2     = R2 + S1
                     X(I2)  = R1*CC1 - S2*SS1
                     Y(I2)  = S2*CC1 - R1*SS1
                     X(I3)  = S3*CC3 + R2*SS3
                     Y(I3)  = R2*CC3 - S3*SS3
30             CONTINUE
               IS = 2*ID - N2 + J
               ID = 4*ID
               IF (IS.LT.N) GOTO 40
20         CONTINUE
10     CONTINUE
```

```
C--------------------SPECIAL LAST STAGE--------------------
      IS = 1
      ID = 4
 50   DO 60 I0 = IS, N, ID
         I1    = I0 + 1
         R1    = X(I0)
         X(I0) = R1 + X(I1)
         X(I1) = R1 - X(I1)
         R1    = Y(I0)
         Y(I0) = R1 + Y(I1)
         Y(I1) = R1 - Y(I1)
 60   CONTINUE
         IS = 2*ID - 1
         ID = 4*ID
      IF (IS.LT.N) GOTO 50
C
C--------------------DIGIT REVERSE COUNTER--------------------
C
 100  J = 1
      N1 = N - 1
      DO 104 I = 1, N1
         IF (I.GE.J) GOTO 101
         XT = X(J)
         X(J) = X(I)
         X(I) = XT
         XT   = Y(J)
         Y(J) = Y(I)
         Y(I) = XT
 101     K = N/2
 102     IF (K.GE.J) GOTO 103
            J = J - K
            K = K/2
            GOTO 102
 103  J = J + K
 104     CONTINUE
         RETURN
         END
```

```
C
C   MAIN PROGRAM: DFTTEST
C   C. S. BURRUS, RICE UNIVERSITY,  JUNE 1983
C
C   TEST OF DIRECT  DFT  CALCULATION
C
      REAL X(60), Y(60), A1(60), B1(60), A(60), B(60), C(60), S(60)
      COMPLEX AC, BC, DC, W
C
C--------------------------INPUT LENGTH----------
  7   WRITE (6,301)
      READ (5,*) N
C
C---------------------------CALULATE CHIRP SIGNAL-----------
          AC = (.9, .3)
      DO 11 K=1,N
          BC = AC**(K-1)
          X(K) = REAL(BC)
          Y(K) = AIMAG(BC)
   11 CONTINUE
C
C----------------------CALCULATE CHIRP DFT---------
C
          TPION = 6.283185308/FLOAT(N)
          W = CMPLX(COS(TPION), -SIN(TPION))
          DC = (1., 0.) - AC**N
      DO 12 K=1,N
          BC = DC/((1.,0.) - AC*W**(K-1))
          A1(K) = REAL(BC)
          B1(K) = AIMAG(BC)
   12 CONTINUE
C
C--------------------DO THE FFT------------------
      CALL DFT(X,Y,A,B,N)
C
C-------------------CALCULATE MAX ERROR-----------
          AA = 0
      DO 14 K=1, N
          AB = ABS(A(K)-A1(K)) + ABS(B(K)-B1(K))
          IF ( AA.LT.AB )  AA = AB
   14 CONTINUE
C
          WRITE (6, 200) AA
      IF (N.NE.1) GOTO 7
  200 FORMAT (/ ' MAX ERROR = ', E14.4)
  301 FORMAT (/ ' ENTER N (N=1 to stop) ')
C
          STOP
      END
```

```
C
C   MAIN PROGRAM: DFTTEST
C   C. S. BURRUS, RICE UNIVERSITY,  JUNE 1983
C
C   TEST OF DFT
C
C   REAL IMPULSE INPUT, COMPLEX OUTPUT
C
C---------------------------------------------------------------
C
        REAL X(260), Y(260),A(260),B(260),C(260),S(260)
C
C----------------------------INPUT LENGTH -------------------
  7     WRITE (6,301)
        READ (5,*) N
C
C----------------------------CALULATE INPUT SIGNAL-----------
C
        WRITE (6,302) N
        DO 100 I1 = 1, N
           DO 200 K1 = 1, N
              X(K1) = 0
  200      CONTINUE
           X(I1) = 1
C
C----------------------DO THE DFT-------------------
C
        CALL DFT(X,Y,A,B,N)
        WRITE (6, 303) I1-1
        WRITE (6, 304)(A(K),K=1,N)
        WRITE (6, 304)(B(K),K=1,N)
C
C--------------------CALCULATE OUTPUT-------------
C
        ER = 0
        WRITE (6, 305)
           DO 300 K = 1, N
              R = COS(FLOAT((K-1)*(I1-1))*6.283185307/FLOAT(N))
              T =-SIN(FLOAT((K-1)*(I1-1))*6.283185307/FLOAT(N))
              ERR = ABS(A(K)-R) + ABS(B(K)-T)
              IF (ER.LT.ERR) ER = ERR
              A(K) = R
  300         B(K) = T
           WRITE (6, 304)(A(K),K=1,N)
           WRITE (6, 304)(B(K),K=1,N)
           WRITE (6, 307) ER
           WRITE (6, 306)
           READ (5,*) L
           IF (L.EQ.0) GOTO 8
           IF (L.EQ.2) GOTO 7
  100   CONTINUE
        GOTO 7
  301   FORMAT (/ ' ENTER N ')
  302   FORMAT ( ' LENGTH = ', I5)
  303   FORMAT ( ' DFT OUTPUT FOR INPUT IMPULSE AT  n = ',I5)
  304   FORMAT ( 7F11.6 )
  305   FORMAT ( ' DFT: REAL, IMAG ')
  306   FORMAT ( ' ENTER 0 TO STOP PROG, 1 FOR NEXT IN, 2 FOR NEW N')
  307   FORMAT ( ' MAX ERROR = ', E14.4)
C
   8       STOP
        END
```

Chapter 5

TMS32010 ASSEMBLY LANGUAGE PROGRAMS FOR THE DFT AND CONVOLUTION

5.1 INTRODUCTION

The TMS32010 assembly language programs in this chapter are based on the FORTRAN routines from Chapter 4. As much as possible, these programs retain the same structure and variable names of their FORTRAN counterparts. Many of the comments in these programs are lines of FORTRAN from the prototype routine which help explain the assembly language.

The FORTRAN programs have been followed as much as possible to retain clarity and consistency. At the same time, some changes have been made to take advantage of the features of the TMS32010 instruction set [1]. First, many of the assembly language programs assume a minimal amount of external memory used to hold the data which is addressed via peripheral I/O instructions, i.e., an address is initially sent to a data address counter and then the data word read from or written to the memory. A counter is used to address the data locations sequentially since complex data are used whose structure consists of the real and imaginary parts of a number in consecutive memory locations. A Q15 format [1] for number representation is assumed for all data and coefficient values. This format (one sign bit + 15 fractional bits; the absolute value of all represented numbers is less than one) simplifies calculations in fixed-point machines such as the TMS32010 digital signal processor. In order to keep the code as easy to read as possible, no scaling was done on intermediate values in any of the TMS32010 programs in this book. These programs can be used by limiting the size of the input to prevent overflow or scaling can easily be added in a conventional way between each stage of the FFT without any speed penalty. Table lookup of coefficients is used for speed and simplicity. These programs use a full size table (often the size of the transform). Smaller tables could be used at the expense of speed if program memory size is limited. All programs have been assembled using a Texas Instruments cross-assembler running on a VAX 11/780 under the VMS operating system.

5.2 TMS32010 ASSEMBLY LANGUAGE PROGRAMS

Program 1 is a direct DFT similar to the first FORTRAN program. All data is initially in on-chip data RAM, and results are output sequentially. The data is on page 0 and is addressed indirectly with the auxiliary registers. All other variables are located on page

1 of the TMS32010. Full 32-bit intermediate values are retained for maximum accuracy. Note that the 32-bit running sum is initialized by shifting the first data point 15 bits which simulates a multiply by one in Q15 format. In this and all the other assembly language programs, arrays and loop counters start at zero, unlike the FORTRAN routines.

Program 2 is a TMS32010 implementation of a second-order Goertzel algorithm with forward-order inputs similar to the FORTRAN Program 4. Like the DFT program, input data is initially on page 0 of data RAM and is addressed indirectly. Results are output sequentially. The 2*cosine factor in the inner loop is implemented by adding twice the product of a single cosine multiply. This 2*cosine term along with the feedback loop makes this algorithm susceptible to quantization error and overflow on a fixed-point machine. Since no scaling is done in the program, the data should be scaled previously to guard against overflow.

A single butterfly radix-2 Cooley-Tukey FFT, similar to FORTRAN Program 6, is shown in Program 3. All data is in external data memory using hardware described in the introduction to this chapter. Because the real and imaginary parts of the complex input data are in sequential locations, not in separate arrays, the data index I has a value twice that in the corresponding FORTRAN routine. The variable IE, the index increment for the coefficients, is computed with multiplies by 2 (implemented with a shift) in order to eliminate divisions which are inefficient. The size of the transform is limited to a power of 2, and the largest transform is simply a function of the amount of available program memory. Program 4 is just like Program 3 but with a single radix-4 butterfly. The transform size is limited to a power of 4.

Program 5 is a radix-4 algorithm with three butterflies, similar to FORTRAN Program 13. Both auxiliary registers are used as loop counters in the program. Auxiliary Register 1 (AR1) replaces K in the DO 10 loop, and Auxiliary Register 0 (AR0) is used in the DO 1 and DO 20 loops. These registers are efficient as counters in code with many loops.

Program 6 is a 64-point complex FFT optimized for execution speed. The algorithm is a three-butterfly radix-4 FFT with all straight-line code. This program uses the macro capability of the assembler to generate the code, which is almost 2800 lines long when expanded. Three macros are used, one for each type of butterfly. All data begins and ends on page 0 of data RAM, and page 1 contains two temporary locations which are addressed by auxiliary registers. The 13-bit immediate coefficients are used as part of the MPYK (multiply immediate) instruction. Butterflies are accessed in the same order as in Program 5 and all data points are addressed directly for speed.

A radix-8 FFT is shown in Program 7. This is a two-butterfly FFT and corresponds to Program 16 in FORTRAN. Because this is a radix-8 program, there are half as many external data accesses as in the looped radix-4 program. This helps to speed up the execution. The tradeoff is in a more complex program and fewer transform sizes which can be computed.

Program 8 is a prime-factor FFT, corresponding to FORTRAN Program 17. This program is specific for a length-63 PFA, but the structure can be used for all module and transform sizes. Coefficients are in a table initially and then read into data RAM for use during execution. Because the unscrambling is not an in-place algorithm, the final unscrambled results are written to port 2, which corresponds to arrays A and B in the FORTRAN programs.

A direct calculation of linear convolution is shown in Program 9. In this example, an infinite input sequence is convolved with a 32-bit permanent sequence which represents the impulse response of a bandpass filter. Input is read sequentially from port 0, and output is written to port 1. The main computation in the convolution is done using the pair of instructions, LTD and MPY, which multiplies the sequences point by point and shifts the input sequence for the next loop. These instructions are especially efficient for this type of inner-product calculation.

5.3 REFERENCE

1. TMS32010 User's Guide (Revision A). Houston, TX: Texas Instruments, 1983.

5.4 SAMPLE TMS32010 PROGRAMS

Table 5-1 lists the nine sample assembly language programs for the TMS32010 that are presented in the succeeding pages.

TABLE 5-1. TMS32010 ASSEMBLY LANGUAGE PROGRAM INDEX

PROGRAM	DESCRIPTION	PAGE
1	Direct DFT with table lookup	148
2	Second-order Goertzel with forward-order input	152
3	Basic one-butterfly Cooley-Tukey radix-2 FFT	156
4	Basic one-butterfly radix-4 FFT	161
5	Three-butterfly radix-4 FFT	168
6	Straight-line 64-point radix-4 FFT	179
7	Two-butterfly radix-8 FFT	190
8	Prime-factor algorithm FFT with unscrambling	205
9	Direct convolution calculation	220

```
          IDT 'DFT'
*
* This program calculates a complex DFT.
*
*         Standard straight DFT algorithm.
*         Uses table lookup of coefficients with table size N
*           for sines and cosines.
*         Maximum size of DFT is 64 points, complex.
*         Keeps a full 32-bit running sum.
*         All data begins in RAM and results are output
*           sequentially to port 0.
*
N EQU 64             * Size of the transform is N.
*
X      EQU 0         * Data points are on page 0.
* Rest of the variables are page 1 locations.
ONE    EQU 0         * Contains value 1
K      EQU 1         * Modulo counter, index into sine table
J      EQU 2         * Count of DFT point
HOLDN  EQU 3         * Contains value N
QUARTN EQU 4         * Contains value N/4
COS    EQU 5         * Current cosine value
SIN    EQU 6         * Current sine value
TABLE  EQU 7         * Contains location of sine table
SUMREH EQU 8         * Current real DFT summation high
SUMREL EQU 9         * Current real DFT summation low
SUMIMH EQU 10        * Current imaginary DFT summation high
SUMIML EQU 11        * Current imaginary DFT summation high
*
          AORG 0
START     LDPK 1
          LACK 1
          SACL ONE
          LT   ONE
          MPYK SINE
          PAC
          SACL TABLE             * Save sine table address
          LACK N
          SACL HOLDN             * HOLDN = N
          LACK N/4
          SACL QUARTN            * QUARTN = N/4
          ZAC
          SACL J                 * J = 0
JLOOP        LARK AR0,N-2
             LARK AR1,0
             LARP 1
             LAC  *+,15
             SACH SUMREH         * Sum real = X(0)
             SACL SUMREL
             LAC  *+,15
             SACH SUMIMH         * Sum imag = Y(0)
             SACL SUMIML
             ZAC
             SACL K              * K = 0
*
ILOOP           LARP 1
                LAC  K
```

```
                ADD  J
                SUB  HOLDN              * K = (K + J)mod N
                BGEZ GT2PI
                   ADD  HOLDN
GT2PI           SACL K                  * 0 <= K < 2PI
                ADD  TABLE
                TBLR SIN                * Get coefficients
                ADD  QUARTN
                TBLR COS
*
* Main DFT calculations.
*
                ZALH SUMREH      ** Real **
                ADDS SUMREL
                LT   COS
                MPY *+
                LTA SIN
                MPY *-
                APAC
                SACH SUMREH      * Sum = sum + cos(k)*x(i) + sin(k)*y(i)
                SACL SUMREL
                ZALH SUMIMH      ** Imaginary **
                ADDS SUMIML
                MPY *+
                SPAC
                LT   COS
                MPY *+,AR0
                APAC
                SACH SUMIMH      * Sum = sum + cos(k)*y(i) - sin(k)*x(i)
                SACL SUMIML
                BANZ ILOOP       * Repeat for i = 2 to N
*
* Output DFT results.
*
             OUT  SUMREH,PA0     * Output real part
             OUT  SUMIMH,PA0     * Output imaginary part
             LAC  J
             ADD  ONE            * J = J + 1
             SACL J
             SUB  HOLDN
             BLZ  JLOOP          * Repeat for J = 1 to N
*
* End of program.
*
STOP    B STOP
*
*  Sine and cosine tables (length N).
*
SINE EQU $
      DATA         0
      DATA      3211
      DATA      6392
      DATA      9511
      DATA     12539
      DATA     15446
      DATA     18204
      DATA     20787
```

```
        DATA      23169
        DATA      25329
        DATA      27244
        DATA      28897
        DATA      30272
        DATA      31356
        DATA      32137
        DATA      32609
COSINE EQU $
        DATA      32767
        DATA      32609
        DATA      32137
        DATA      31356
        DATA      30272
        DATA      28897
        DATA      27244
        DATA      25329
        DATA      23169
        DATA      20787
        DATA      18204
        DATA      15446
        DATA      12539
        DATA       9511
        DATA       6392
        DATA       3211
        DATA          0
        DATA      -3211
        DATA      -6392
        DATA      -9511
        DATA     -12539
        DATA     -15446
        DATA     -18204
        DATA     -20787
        DATA     -23169
        DATA     -25329
        DATA     -27244
        DATA     -28897
        DATA     -30272
        DATA     -31356
        DATA     -32137
        DATA     -32609
        DATA     -32767
        DATA     -32609
        DATA     -32137
        DATA     -31356
        DATA     -30272
        DATA     -28897
        DATA     -27244
        DATA     -25329
        DATA     -23169
        DATA     -20787
        DATA     -18204
        DATA     -15446
        DATA     -12539
        DATA      -9511
        DATA      -6392
        DATA      -3211
```

```
DATA        0
DATA     3211
DATA     6392
DATA     9511
DATA    12539
DATA    15446
DATA    18204
DATA    20787
DATA    23169
DATA    25329
DATA    27244
DATA    28897
DATA    30272
DATA    31356
DATA    32137
DATA    32609
END
```

```
          IDT 'GZEL'
*
*  Program for computation of DFT using Goertzel's algorithm.
*
*       Uses second-order Goertzel's algorithm.
*       Computes up to 64 DFT points.
*       Data points are located on page 0.
*       Results are output sequentially to port 0.
*       Coefficient table is of size N for sine and cosines.
*
N EQU 64                  * Size of the transform
          AORG 0
*
*
X       EQU 0        * Data points
* Other locations on page 1.
ONE     EQU 0        * Contains value 1
HOLDN   EQU 1        * Contains value N
QUARTN  EQU 2        * Contains value N/4
J       EQU 3        * Current DFT point
COS     EQU 4        * Current cosine value
SIN     EQU 5        * Current sine value
TABLE   EQU 6        * Location of sine table
A1      EQU 7        * Real locations
A2      EQU 8
B1      EQU 9        * Imaginary locations
B2      EQU 10
REAL    EQU 11       * Real result for output
IMAG    EQU 12       * Imaginary result for output
*
* Start of routine.
*
START     LDPK 1
          LACK 1
          SACL ONE
          LT   ONE
          MPYK SINE
          PAC
          SACL TABLE            * Save address of sine table
          LACK N
          SACL HOLDN            * HOLDN = N
          LACK N/4
          SACL QUARTN           * QUARTN = N/4
          ZAC
          SACL J                * J = 0
*
LOOP            LAC  TABLE
                ADD  J
                TBLR SIN        * Get sine and cosine values
                ADD  QUARTN     * for this loop
                TBLR COS
*****************************************************************
*                                                               *
*         MAIN LOOP FOR GOERTZEL'S ALGORITHM                    *
*                                                               *
*****************************************************************
GOTCOS       ZAC
```

```
              SACL A2              * A2 = 0
              SACL B2              * B2 = 0
              LARP 1
              LARK AR0,N-2
              LARK AR1,X
              LAC  *+
              SACL A1              * A1 = X(0)
              LAC  *+
              SACL B1              * B1 = Y(0)
              LT   COS
NLOOP            LARP 1
                 MPY  A1
                 PAC
                 APAC              * ACC = 2cos(j)*A1
                 SUB  A2,15
                 ADD  *+,15        * ACC = 2cos(j)*A1 - A2 + X(i)
                 DMOV A1           * A2 = A1
                 SACH A1,1         * A1 = 2cos(j)*A1 - A2 + X(i)
                 MPY  B1
                 PAC
                 APAC              * ACC = 2cos(j)*B1
                 SUB  B2,15
                 ADD  *+,15,AR0    * ACC = 2cos(j)*B1 - B2 + Y(i)
                 DMOV B1           * B2 = B1
                 SACH B1,1         * B1 = 2cos(j)*B1 - B2 + Y(i)
                 BANZ NLOOP        * Loop for i = 1 to N-1
*
* Main loop complete.
*
              MPY  A1
              PAC
              SUB  A2,15
              LT   SIN
              MPY  B1
              SPAC
              SACH REAL,1          * Real = cos(j)*A1 - A2 - sin(j)*B1
              MPY  A1
              PAC
              LT   COS
              MPY  B1
              APAC
              SUB  B2,15
              SACH IMAG,1          * Imag = cos(j)*B1 - B2 + sin(j)*A1
              OUT  REAL,PA0        * Output results
              OUT  IMAG,PA0
              LAC  J
              ADD  ONE
              SACL J               * J = J + 1
              SUB  HOLDN
              BLZ  LOOP            * Repeat for J = 0 to N-1
*
* End of program.
*
STOP   B STOP
*
* Sine and cosine tables.
*
```

```
SINE   EQU $
       DATA 0
       DATA 3211
       DATA 6392
       DATA 9511
       DATA 12539
       DATA 15446
       DATA 18204
       DATA 20787
       DATA 23169
       DATA 25329
       DATA 27244
       DATA 28897
       DATA 30272
       DATA 31356
       DATA 32137
       DATA 32609
COSINE EQU $
       DATA 32767
       DATA 32609
       DATA 32137
       DATA 31356
       DATA 30272
       DATA 28897
       DATA 27244
       DATA 25329
       DATA 23169
       DATA 20787
       DATA 18204
       DATA 15446
       DATA 12539
       DATA 9511
       DATA 6392
       DATA 3211
       DATA 0
       DATA -3211
       DATA -6392
       DATA -9511
       DATA -12539
       DATA -15446
       DATA -18204
       DATA -20787
       DATA -23169
       DATA -25329
       DATA -27244
       DATA -28897
       DATA -30272
       DATA -31356
       DATA -32137
       DATA -32609
       DATA -32767
       DATA -32609
       DATA -32137
       DATA -31356
       DATA -30272
       DATA -28897
       DATA -27244
```

```
DATA -25329
DATA -23169
DATA -20787
DATA -18204
DATA -15446
DATA -12539
DATA -9511
DATA -6392
DATA -3211
DATA 0
DATA 3211
DATA 6392
DATA 9511
DATA 12539
DATA 15446
DATA 18204
DATA 20787
DATA 23169
DATA 25329
DATA 27244
DATA 28897
DATA 30272
DATA 31356
DATA 32137
DATA 32609
END
```

```
        IDT 'FFT2'
*
*    Cooley-Tukey Radix-2, DIF FFT Program for the TMS32010.
*
*        Single FFT butterfly.
*        Complex input data - size limited only by program memory availability.
*        Uses table lookup of the twiddle factors.
*        No scaling is done on intermediate values in the program.
*        External data RAM is addressed via peripheral I/O instructions.
*             An address counter is required and is loaded by a write to
*             port 0.  Data is read from and written to RAM via port 1.
*             The address counter should increment after every read or write.
*        Data in the external RAM assumes complex data with corresponding
*             real and imaginary data values in consecutive locations.
*
*
* N is the size of the transform.  N = 2**M.
N EQU 64
M EQU 6
*
* Data Memory Allocation.
*
XI       EQU 0             * Array value X(I)
YI       EQU 1             * Array value Y(I)
XL       EQU 2             * Array value X(L)
YL       EQU 3             * Array value Y(L)
XT       EQU 4             * Temporary - real part
YT       EQU 5             * Temporary - imaginary part
I        EQU 6             * 1st index
L        EQU 7             * 2nd index
COS      EQU 8             * Twiddle factor - real part
SIN      EQU 9             * Twiddle factor - imaginary part
IA       EQU 10            * Index to twiddle factors
IE       EQU 11            * Increment to IA
HOLDN    EQU 12            * Contains value N
QUARTN   EQU 13            * Contains value N/4
N1       EQU 14            * Increment to I.
N2       EQU 15            * Separation of I and L
J        EQU 16            * Loop counter
ONE      EQU 17            * Contains value 1
TABLE    EQU 18            * Location of coefficient table
*
* Begin program memory section.
*
         AORG 0
START    LDPK 0
         LACK 1
         SACL ONE          * Initialize IE = 1
         SACL IE
         LT   ONE
         MPYK SINE
         PAC
         SACL TABLE        * Table has address of cosine table
         MPYK N
         PAC
         SACL HOLDN        * Holdn = N
         SACL N2           * Initialize N2 = N
```

```
            LAC   HOLDN,14
            SACH QUARTN           * Quartn = N/4
            LARK AR0,M-1          * AR0 contains K counter
KLOOP          LARP 1
               LAC  N2,15
               SACH N1,1          * N1 = N2
               SACH N2            * N2 = N2/2
               ZAC
               SACL IA
               SACL J
               LAR  AR1,N2        * AR1 contains J value
               MAR  *-            * Start at N2-1
JLOOP             LAC  TABLE      * Table is full size
                  ADD  IA
                  TBLR SIN         * Get twiddle factors
                  ADD  QUARTN
                  TBLR COS
                  LAC  IA
                  ADD  IE
                  SACL IA         * IA =  IA + IE
                  LAC  J,1
                  SACL I          * I = J  (data organized as real value followed
*                                 * by imaginary so address I is 2 times J).
ILOOP                LAC  I
                     ADD  N2,1              * L = I + N2
                     SACL L
*
                     OUT  I,PA0             * Output address of XI
                     IN   XI,PA1            * Read real and imaginary parts
                     IN   YI,PA1
                     OUT  L,PA0             * Output address of XL
                     IN   XL,PA1            * Read real and imaginary parts
                     IN   YL,PA1
*
* Compute butterfly.
*
                     LAC  XI
                     SUB  XL
                     SACL XT                * XT = XI - XL
                     ADD  XL,1
                     SACL XI                * XI = XI + XL
                     LAC  YI
                     SUB  YL
                     SACL YT                * YT = YI - YL
                     ADD  YL,1
                     SACL YI                * YI = YI + YL
                     LT   COS
                     MPY  YT
                     PAC
                     LT   SIN
                     MPY  XT
                     SPAC
                     SACH YL,1              * YL = COS*YT - SIN*XT
                     MPY  YT
                     PAC
                     LT   COS
                     MPY  XT
```

```
                    APAC
                    SACH XL,1              * XL = COS*XT + SIN*YT
*
* Output results of the butterfly.
*
                    OUT  I,PA0             * Output I value address.
                    OUT  XI,PA1            * Output real and imaginary parts.
                    OUT  YI,PA1
                    OUT  L,PA0             * Output L value address.
                    OUT  XL,PA1            * Output real and imaginary parts.
                    OUT  YL,PA1
*
* Add increment for next loop.
*
                    LAC  I
                    ADD  N1,1              * I = I + N1
                    SACL I
                    SUB  HOLDN,1           * While I < N
                    BLZ  ILOOP
                LAC  J
                ADD  ONE                   * J = J + 1
                SACL J
                BANZ JLOOP
             LAC  IE,1
             SACL IE                       * IE = 2 * IE
             LARP 0
             BANZ KLOOP
*
* Digit reverse counter for radix-2 FFT computation.
*
DRC2    ZAC
        SACL L
        SACL I
        LARP 0
        LAR  AR0,HOLDN                     * For I = 0 to N-2
        MAR  *-
        MAR  *-
DRLOOP          SUB  L                     * If I < L, then swap
                BGEZ NOSWAP
* Swap i and l values.
                        OUT  I,PA0
                        IN   XI,PA1
                        IN   YI,PA1
                        OUT  L,PA0
                        IN   XL,PA1
                        IN   YL,PA1
                        OUT  L,PA0
                        OUT  XI,PA1
                        OUT  YI,PA1
                        OUT  I,PA0
                        OUT  XL,PA1
                        OUT  YL,PA1
NOSWAP          LAC  HOLDN
                SACL J                            * J = N
INLOOP                  LAC  L
                        SUB  J                    * If L >= J then
                        BLZ  OUTL
```

```
                                  SACL L          * L = L - J
                                  LAC  J,15
                                  SACH J          * J = J/2.
                                  B    INLOOP
OUTL            ADD  J,1
                SACL L                            * L = L + J
                LAC  I
                ADD  ONE,1
                SACL I                            * Increment I
                BANZ DRLOOP
*
*  FFT complete.
*
WHOA    B WHOA
*
* Coefficient table (size of table is 3n/4).
*
SINE EQU $
      DATA 0
      DATA 3211
      DATA 6392
      DATA 9511
      DATA 12539
      DATA 15446
      DATA 18204
      DATA 20787
      DATA 23169
      DATA 25329
      DATA 27244
      DATA 28897
      DATA 30272
      DATA 31356
      DATA 32137
      DATA 32609
COSINE EQU $
      DATA 32767
      DATA 32609
      DATA 32137
      DATA 31356
      DATA 30272
      DATA 28897
      DATA 27244
      DATA 25329
      DATA 23169
      DATA 20787
      DATA 18204
      DATA 15446
      DATA 12539
      DATA 9511
      DATA 6392
      DATA 3211
      DATA 0
      DATA -3211
      DATA -6392
      DATA -9511
      DATA -12539
      DATA -15446
```

```
        DATA -18204
        DATA -20787
        DATA -23169
        DATA -25329
        DATA -27244
        DATA -28897
        DATA -30272
        DATA -31356
        DATA -32137
        DATA -32609
        END
```

```
          IDT 'FFT4'
*
*     Cooley-Tukey Radix-4, DIF FFT Program
*
*        Single radix-4 butterfly.
*        Complex input data - size limited only by program memory availability.
*        Uses table lookup of twiddle factors.
*        No scaling is done on intermediate values in the program.
*        External data RAM is addressed via peripheral I/O instructions.
*                An address counter is required and is loaded by a write
*                to port 0.  Data is read from and written to port 1.
*        Data in the external RAM assumes complex data with corresponding
*                real and imaginary data values in consecutive locations.
*
* N is the size of the transform (N = 4**M).
N EQU 64
M EQU 3
*
* Data memory allocation.
*
XI      EQU 0              * Data values for butterfly
YI      EQU 1
XI1     EQU 2
YI1     EQU 3
XI2     EQU 4
YI2     EQU 5
XI3     EQU 6
YI3     EQU 7
I       EQU 8              * Data indices
I1      EQU 9
I2      EQU 10
I3      EQU 11
CO1     EQU 12             * Twiddle factor coefficients
CO2     EQU 13
CO3     EQU 14
SI1     EQU 15
SI2     EQU 16
SI3     EQU 17
N1      EQU 18             * Increment to I
N2      EQU 19             * Index separation
IA      EQU 20             * Index to twiddle factors
IE      EQU 21             * Increment to IA
J       EQU 22             * Counters
K       EQU 23
R1      EQU 24             * Temporaries
R2      EQU 25
S1      EQU 26
S2      EQU 27
R3      EQU 28
R4      EQU 29
S3      EQU 30
S4      EQU 31
XJ      EQU 32
YJ      EQU 33
TEMP    EQU 34
HOLDN   EQU 35             * Contains the value N
QUARTN  EQU 36             * Contains the value N/4
```

```
ONE     EQU 37              * Contains the value 1
TABLE   EQU 38              * Location of coefficient table
*
* Begin program memory section.
*
          AORG 0
START     LDPK 0
          LACK 1
          SACL IE              * Initialize IE = 1
          SACL ONE
          LT   ONE
          MPYK SINE
          PAC
          SACL TABLE           * Save address of coefficient table
          MPYK N
          PAC
          SACL HOLDN
          SACL N2              * Initialize N2 = N
          LAC  HOLDN,14
          SACH QUARTN          * QUARTN = N/4
          LARK AR0,M-1         * AR0 contains K value
KLOOP        LARP 1
             LAC  N2
             SACL N1           * N1 = N2
             LAC  N2,14
             SACH N2           * N2 = N2/4
             ZAC
             SACL J
             SACL IA
             LAR  AR1,N2       * AR1 contains J value
             MAR  *-           * Start at N2-1
JLOOP           LAC  TABLE
                ADD  IA
                TBLR SI1       * Get twiddle factor 1
                ADD  QUARTN
                TBLR CO1
                LAC  TABLE
                ADD  IA,1
                TBLR SI2       * Get twiddle factor 2
                ADD  QUARTN
                TBLR CO2
                LAC  TABLE
                ADD  IA,1
                ADD  IA
                TBLR SI3       * Get twiddle factor 3
                ADD  QUARTN
                TBLR CO3
                LAC  IA
                ADD  IE
                SACL IA          * IA = IA + IE
                LAC  J,1
                SACL I           * I = J (data organized ar real value followed
*                                * by imaginary so address I is 2 times J).
ILOOP              LAC  I
                   ADD  N2,1
                   SACL I1            * I1 = I + N2
                   ADD  N2,1
```

```
                SACL I2                 * I2 = I1 + N2
                ADD  N2,1
                SACL I3                 * I3 = I2 + N2
*
                OUT  I,PA0              * Output address of XI
                IN   XI,PA1             * Read real and imaginary parts
                IN   YI,PA1
                OUT  I1,PA0              * Output address of XI1
                IN   XI1,PA1             * Read real and imaginary parts
                IN   YI1,PA1
                OUT  I2,PA0              * Output address of XI2
                IN   XI2,PA1             * Read real and imaginary parts
                IN   YI2,PA1
                OUT  I3,PA0              * Output address of XI3
                IN   XI3,PA1             * Read real and imaginary parts
                IN   YI3,PA1
*
* Compute butterfly.
*
                LAC  XI
                ADD  XI2
                SACL R1                 * R1 = X(I) + X(I2)
                SUB  XI2,1
                SACL R3                 * R3 = X(I) - X(I2)
                LAC  YI
                ADD  YI2
                SACL S1                 * S1 = Y(I) + Y(I2)
                SUB  YI2,1
                SACL S3                 * S3 = Y(I) - Y(I2)
                LAC  XI1
                ADD  XI3
                SACL R2                 * R2 = X(I1) + X(I3)
                SUB  XI3,1
                SACL R4                 * R4 = X(I1) - X(I3)
                LAC  YI1
                ADD  YI3
                SACL S2                 * S2 = Y(I1) + Y(I3)
                SUB  YI3,1
                SACL S4                 * S4 = Y(I1) - Y(I3)
*
                LAC  R1
                ADD  R2
                SACL XI                 * X(I) = R1 + R2
                SUB  R2,1
                SACL R2                 * R2 = R1 - R2
                LAC  R3
                SUB  S4
                SACL R1                 * R1 = R3 - S4
                ADD  S4,1
                SACL R3                 * R3 = R3 + S4
*
                LAC  S1
                ADD  S2
                SACL YI                 * Y(I) = S1 + S2
                SUB  S2,1
                SACL S2                 * S2 = S1 - S2
                LAC  S3
```

```
                ADD  R4
                SACL S1              * S1 = S3 + R4
                SUB  R4,1
                SACL S3              * S3 = S3 - R4
*
                LT   CO1
                MPY  S3
                PAC
                LT   SI1
                MPY  R3
                SPAC
                SACH YI1,1           * Y(I1) = CO1*S3 - SI1*R3
                MPY  S3
                PAC
                LT   CO1
                MPY  R3
                LTA  CO2
                SACH XI1,1           * X(I1) = CO1*R3 + SI1*S3
                MPY  S2
                PAC
                LT   SI2
                MPY  R2
                SPAC
                SACH YI2,1           * Y(I2) = CO2*S2 - SI2*R2
                MPY  S2
                PAC
                LT   CO2
                MPY  R2
                LTA  CO3
                SACH XI2,1           * X(I2) = CO2*R2 + SI2*S2
                MPY  S1
                PAC
                LT   SI3
                MPY  R1
                SPAC
                SACH YI3,1           * Y(I3) = CO3*S1 - SI3*R1
                MPY  S1
                PAC
                LT   CO3
                MPY  R1
                APAC
                SACH XI3,1           * X(I3) = CO3*R1 + SI3*S1
*
* Output results of the butterfly.
*
                OUT  I,PA0
                OUT  XI,PA1
                OUT  YI,PA1
                OUT  I1,PA0
                OUT  XI1,PA1
                OUT  YI1,PA1
                OUT  I2,PA0
                OUT  XI2,PA1
                OUT  YI2,PA1
                OUT  I3,PA0
                OUT  XI3,PA1
                OUT  YI3,PA1
```

```
*
* Add increment for next loop.
*
                   LAC   I
                   ADD   N1,1
                   SACL  I                  * I = I + N1
                   SUB   HOLDN,1
                   BLZ   ILOOP
               LAC   J
               ADD   ONE
               SACL  J                      * J = J + 1
               BANZ  JLOOP
            LAC   IE,2
            SACL  IE                        * IE = IE*4
            LARP  0
            BANZ  KLOOP
*
* Digit reverse counter for radix-4 FFT computation.
*
DRC4    ZAC
        SACL  J
        SACL  I
        LARP  0
        LAR   AR0,HOLDN
        MAR   *-
        MAR   *-
DRLOOP          SUB   J
                BGEZ  NOSWAP
*
* Swap I and J locations.
*
                        OUT   I,PA0
                        IN    XI,PA1
                        IN    YI,PA1
                        OUT   J,PA0
                        IN    XJ,PA1
                        IN    YJ,PA1
                        OUT   J,PA0
                        OUT   XI,PA1
                        OUT   YI,PA1
                        OUT   I,PA0
                        OUT   XJ,PA1
                        OUT   YJ,PA1
NOSWAP          LAC   QUARTN,1
                SACL  K
INLOOP                  LT    K
                        MPYK  3
                        PAC
                        SACL  TEMP
                        SUB   J
                        BGZ   OUTL
                           LAC   J
                           SUB   TEMP
                           SACL  J
                           LAC   K,14
                           SACH  K
                           B     INLOOP
```

```
OUTL                LAC  J
                    ADD  K
                    SACL J
                    LAC  I
                    ADD  ONE,1
                    SACL I
                    BANZ DRLOOP
*
* End of FFT.
*
STOP  B STOP
*
* Sine and cosine tables for coefficients.
*
SINE EQU $
      DATA 0
      DATA 3211
      DATA 6392
      DATA 9511
      DATA 12539
      DATA 15446
      DATA 18204
      DATA 20787
      DATA 23169
      DATA 25329
      DATA 27244
      DATA 28897
      DATA 30272
      DATA 31356
      DATA 32137
      DATA 32609
COSINE EQU $
      DATA 32767
      DATA 32609
      DATA 32137
      DATA 31356
      DATA 30272
      DATA 28897
      DATA 27244
      DATA 25329
      DATA 23169
      DATA 20787
      DATA 18204
      DATA 15446
      DATA 12539
      DATA 9511
      DATA 6392
      DATA 3211
      DATA 0
      DATA -3211
      DATA -6392
      DATA -9511
      DATA -12539
      DATA -15446
      DATA -18204
      DATA -20787
      DATA -23169
```

```
DATA -25329
DATA -27244
DATA -28897
DATA -30272
DATA -31356
DATA -32137
DATA -32609
DATA -32767
DATA -32609
DATA -32137
DATA -31356
DATA -30272
DATA -28897
DATA -27244
DATA -25329
DATA -23169
DATA -20787
DATA -18204
DATA -15446
DATA -12539
DATA -9511
DATA -6392
DATA -3211
END
```

```
         IDT 'FFT4-3'
*
*    Cooley-Tukey Radix-4, DIF FFT Program
*
*       Three radix-4 butterflies.
*       Complex input data - size limited only by program memory availability.
*       Uses table lookup of twiddle factors.
*       No scaling is done on intermediate values in the program.
*       External data RAM is addressed via peripheral I/O instructions.
*               An address counter is required and is loaded by a write
*               to port 0.  Data is read from and written to port 1.
*       Data in the external RAM assumes complex data with corresponding
*               real and imaginary data values in consecutive locations.
*
* N is the size of the transform (N = 4**M).
N EQU 64
M EQU 3
*
* Data memory allocation.
*
XI      EQU 0                  * Data values for butterfly
YI      EQU 1
XI1     EQU 2
YI1     EQU 3
XI2     EQU 4
YI2     EQU 5
XI3     EQU 6
YI3     EQU 7
I       EQU 8                  * Data indices
I1      EQU 9
I2      EQU 10
I3      EQU 11
CO1     EQU 12                 * Twiddle factor coefficients
CO2     EQU 13
CO3     EQU 14
SI1     EQU 15
SI2     EQU 16
SI3     EQU 17
N1      EQU 18                 * Increment to I
N2      EQU 19                 * Index separation
IA      EQU 20                 * Index to twiddle factors
IE      EQU 21                 * Increment to IA
J       EQU 22                 * Counters
K       EQU 23
R1      EQU 24                 * Temporaries
R2      EQU 25
S1      EQU 26
S2      EQU 27
R3      EQU 28
R4      EQU 29
S3      EQU 30
S4      EQU 31
XJ      EQU 32
YJ      EQU 33
TEMP    EQU 34
T       EQU 35
HOLDN   EQU 36                 * Contains the value N
```

```
QUARTN EQU 37           * Contains the value N/4
ONE    EQU 38           * Contains the value 1
TABLE  EQU 39           * Location of coefficient table
JT     EQU 40
C21    EQU 41           * Contains value .707106778
*
* Begin program memory section.
*
          AORG 0
START     LDPK 0
          LACK 1
          SACL IE              * Initialize IE = 1
          SACL ONE
          LT   ONE
          MPYK SINE
          PAC
          SACL TABLE           * Save location of sine table
          SUB  ONE
          TBLR C21             * Read coefficient for W = J
          MPYK N
          PAC
          SACL HOLDN
          SACL N2              * Initialize N2 = N
          LAC  HOLDN,14
          SACH QUARTN          * QUARTN = N/4
          LARK AR1,M-1         * AR0 contains K value
KLOOP        LAC  N2
             SACL N1           * N1 = N2
             LAC  N2,13
             SACH N2,1         * N2 = N2/4
             SACH JT           * JT = N2/2
*********************************************************************
*
*  Special butterfly for theta = 0.
*
*********************************************************************
            LARP 0
            LAR  AR0,IE          * Loop counter - repeat loop IE times
            MAR  *-
            ZAC
            SACL I               * I = 0.
ZERO        LAC  I
            ADD  N2,1
            SACL I1              * I1 = I + N2
            ADD  N2,1
            SACL I2              * I2 = I1 + N2
            ADD  N2,1
            SACL I3              * I3 = I2 + N2
*
* Read in four data values.
*
            OUT  I,PA0
            IN   XI,PA1
            IN   YI,PA1
            OUT  I1,PA0
            IN   XI1,PA1
            IN   YI1,PA1
```

```
          OUT  I2,PA0
          IN   XI2,PA1
          IN   YI2,PA1
          OUT  I3,PA0
          IN   XI3,PA1
          IN   YI3,PA1
*
* Start butterfly.
*
          LAC  XI
          ADD  XI2
          SACL R1                * R1 = X(I) + X(I2)
          SUB  XI2,1
          SACL R2                * R2 = X(I) - X(I2)
          LAC  XI1
          ADD  XI3
          SACL R3                * R3 = X(I1) + X(I3)
          ADD  R1
          SACL XI                * X(I) = R1 + R3
          SUB  R3,1
          SACL XI2               * X(I2) = R1 - R3
*
          LAC  YI
          ADD  YI2
          SACL R1                * R1 = Y(I) + Y(I2)
          SUB  YI2,1
          SACL R4                * R4 = Y(I) - Y(I2)
          LAC  YI1
          ADD  YI3
          SACL R3                * R3 = Y(I1) + Y(I3)
          ADD  R1
          SACL YI                * Y(I) = R1 + R3
          SUB  R3,1
          SACL YI2               * Y(I2) = R1 - R3
*
          LAC  XI1
          SUB  XI3
          SACL R1                * R1 = X(I1) - X(I3)
          LAC  YI1
          SUB  YI3
          SACL R3                * R3 = Y(I1) - Y(I3)
          ADD  R2
          SACL XI1               * X(I1) = R2 + R3
          SUB  R3,1
          SACL XI3               * X(I3) = R2 - R3
          LAC  R4
          SUB  R1
          SACL YI1               * Y(I1) = R4 - R1
          ADD  R1,1
          SACL YI3               * Y(I3) = R4 + R1
*
* Output four results.
*
          OUT  I,PA0
          OUT  XI,PA1
          OUT  YI,PA1
          OUT  I1,PA0
```

```
            OUT  XI1,PA1
            OUT  YI1,PA1
            OUT  I2,PA0
            OUT  XI2,PA1
            OUT  YI2,PA1
            OUT  I3,PA0
            OUT  XI3,PA1
            OUT  YI3,PA1
*
            LAC  I
            ADD  N1,1
            SACL I                * I = I + N1
            BANZ ZERO
*
* End of W = 0 butterfly.
*
                LARP 1
                BANZ NORM         * If this is last phase,
                B DRC4            * go to end.
*****************************************************************
*
* Standard radix-4 butterfly.
*
*****************************************************************
NORM        LAC  IE
            SACL IA             * IA = IE
            LARP 0
            LAR  AR0,N2
            MAR  *-
            MAR  *-             * Loop for J = 1 to N2-1
            LACK 1
            SACL J
JLOOP          SUB  JT
               BZ   SPEC        * If J = JT, then use special butterfly.
*
               LAC  TABLE
               ADD  IA
               TBLR SI1         * Get twiddle factor 1
               ADD  QUARTN
               TBLR CO1
               LAC  TABLE
               ADD  IA,1
               TBLR SI2         * Get twiddle factor 2
               ADD  QUARTN
               TBLR CO2
               LAC  TABLE
               ADD  IA,1
               ADD  IA
               TBLR SI3         * Get twiddle factor 3
               ADD  QUARTN
               TBLR CO3
               LAC  IA
               ADD  IE
               SACL IA            * IA = IA + IE
               LAC  J,1
               SACL I             * I = J (data organized as real value followed
*                                 * by imaginary so address I is 2 times J).
```

```
ILOOP               LAC  I
                    ADD  N2,1
                    SACL I1            * I1 = I + N2
                    ADD  N2,1
                    SACL I2            * I2 = I1 + N2
                    ADD  N2,1
                    SACL I3            * I3 = I2 + N2
*
                    OUT  I,PA0         * Output address of XI
                    IN   XI,PA1        * Read real and imaginary parts
                    IN   YI,PA1
                    OUT  I1,PA0         * Output address of XI1
                    IN   XI1,PA1        * Read real and imaginary parts
                    IN   YI1,PA1
                    OUT  I2,PA0         * Output address of XI2
                    IN   XI2,PA1        * Read real and imaginary parts
                    IN   YI2,PA1
                    OUT  I3,PA0         * Output address of XI3
                    IN   XI3,PA1        * Read real and imaginary parts
                    IN   YI3,PA1
*
* Compute butterfly.
*
                    LAC  XI
                    ADD  XI2
                    SACL R1            * R1 = X(I) + X(I2)
                    SUB  XI2,1
                    SACL R2            * R2 = X(I) - X(I2)
                    LAC  XI1
                    ADD  XI3
                    SACL T             * T = X(I1) + X(I3)
                    ADD  R1
                    SACL XI            * X(I) = R1 + T
                    SUB  T,1
                    SACL R1            * R1 = R1 - T
                    LAC  YI
                    ADD  YI2
                    SACL S1            * S1 = Y(I) + Y(I2)
                    SUB  YI2,1
                    SACL S2            * S2 = Y(I) - Y(I2)
                    LAC  YI1
                    ADD  YI3
                    SACL T             * T = Y(I1) + Y(I3)
                    ADD  S1
                    SACL YI            * Y(I) = S1 + T
                    SUB  T,1
                    SACL S1            * S1 = S1 -T
*
                    LT   CO2
                    MPY  S1
                    PAC
                    LT   SI2
                    MPY  R1
                    SPAC
                    SACH YI2,1         * Y(I2) = (S1*CO2) - (R1*SI2)
                    MPY  S1
                    PAC
```

```
                LT   CO2
                MPY  R1
                LTA  CO1
                SACH XI2,1          * X(I2) = (R1*CO2) + (SI*SI2)
*
                LAC  YI1
                SUB  YI3
                SACL T              * T = Y(I1) - Y(I3)
                ADD  R2
                SACL R1             * R1 = R2 + T
                SUB  T,1
                SACL R2             * R2 = R2 - T
                LAC  XI1
                SUB  XI3
                SACL T              * T = X(I1) - X(I3)
                LAC  S2
                SUB  T
                SACL S1             * S1 = S2 - T
                ADD  T,1
                SACL S2             * S2 = S2 + T
***
                MPY  S1
                PAC
                LT   SI1
                MPY  R1
                SPAC
                SACH YI1,1          * Y(I1) = (S1*CO1) - (R1*SI1)
                MPY  S1
                PAC
                LT   CO1
                MPY  R1
                LTA  CO3
                SACH XI1,1          * X(I1) = (R1*CO1) + (S1*SI1)
                MPY  S2
                PAC
                LT   SI3
                MPY  R2
                SPAC
                SACH YI3,1          * Y(I3) = (S2*CO3) - (R2*SI3)
                MPY  S2
                PAC
                LT   CO3
                MPY  R2
                APAC
                SACH XI3,1          * X(I3) = (R2*CO3) + (S2*SI3)
*
* Output results of the butterfly.
*
                OUT  I,PA0
                OUT  XI,PA1
                OUT  YI,PA1
                OUT  I1,PA0
                OUT  XI1,PA1
                OUT  YI1,PA1
                OUT  I2,PA0
                OUT  XI2,PA1
                OUT  YI2,PA1
```

```
                   OUT  I3,PA0
                   OUT  XI3,PA1
                   OUT  YI3,PA1
*
* Add increment for next loop.
*
                   LAC  I
                   ADD  N1,1
                   SACL I                  * I = I + N1
                   SUB  HOLDN,1
                   BLZ  ILOOP              * Loop while I < N
               LAC  J
               ADD  ONE
               SACL J                      * J = J + 1
               BANZ JLOOP
           LAC  IE,2
           SACL IE                         * IE = IE * 4
           B    KLOOP
********************************************************************
*
*  Special routine for theta = pi/4.
*
********************************************************************
SPEC        LAC  J,1
            SACL I                         * I = J
SLOOP       LAC  I
            ADD  N2,1
            SACL I1                        * I1 = I + N2
            ADD  N2,1
            SACL I2                        * I2 = I1 + N2
            ADD  N2,1
            SACL I3                        * I3 = I2 +N2
*
* Input data values.
*
            OUT  I,PA0
            IN   XI,PA1
            IN   YI,PA1
            OUT  I1,PA0
            IN   XI1,PA1
            IN   YI1,PA1
            OUT  I2,PA0
            IN   XI2,PA1
            IN   YI2,PA1
            OUT  I3,PA0
            IN   XI3,PA1
            IN   YI3,PA1
*
* Compute butterfly.
*
            LAC  XI
            ADD  XI2
            SACL R1                * R1 = X(I) + X(I2)
            SUB  XI2,1
            SACL R2                * R2 = X(I) - X(I2)
            LAC  YI
            ADD  YI2
```

```
            SACL S1              * S1 = Y(I) + Y(I2)
            SUB  YI2,1
            SACL S2              * S2 = Y(I) - Y(I2)
            LAC  XI1
            ADD  XI3
*                                * T = X(I1) + X(I3)
            ADD  R1
            SACL XI              * X(I) = T + R1
            SUB  R1,1
            SACL YI2             * Y(I2) = T - R1
            LAC  YI1
            ADD  YI3
            SACL T               * T = Y(I1) + Y(I3)
            ADD  S1
            SACL YI              * Y(I) = S1 + T
            SUB  T,1
            SACL XI2             * X(I2) = S1 - T
            LAC  XI1
            SUB  XI3
            SACL R1              * R1 = X(I1) - X(I3)
            LAC  YI1
            SUB  YI3
            SACL S1              * S1 = Y(I1) - Y(I3)
            ADD  R2
            SACL T               * T = R2 + S1
            SUB  S1,1
            SACL R2              * R2 = R2 - S1
            LAC  S2
            SUB  R1
            SACL S1              * S1 = S2 - R1
            ADD  R1,1
            SACL S2              * S2 = S2 + R1
*
            LT   C21
            MPY  S1
            PAC
            MPY  T
            APAC
            SACH XI1,1           * X(I1) = (T + S1) * C21
            SPAC
            SPAC
            SACH YI1,1           * Y(I1) = (S1 - T) * C21
            ZAC
            MPY  R2
            SPAC
            MPY  S2
            APAC
            SACH XI3,1           * X(I3) = (S2 - R2) * C21
            SPAC
            SPAC
            SACH YI3,1           * Y(I3) = -(S2 + R2) * C21
*
* Output results.
*
            OUT  I,PA0
            OUT  XI,PA1
            OUT  YI,PA1
```

```
              OUT  I1,PA0
              OUT  XI1,PA1
              OUT  YI1,PA1
              OUT  I2,PA0
              OUT  XI2,PA1
              OUT  YI2,PA1
              OUT  I3,PA0
              OUT  XI3,PA1
              OUT  YI3,PA1
*
               LAC  I
               ADD  N1,1
               SACL I                     * I = I + N1
               SUB  HOLDN,1
               BLZ  SLOOP                 * Loop while I < N
            LAC  J
            ADD  ONE
            SACL J                        * J = J + 1
            LAC  IA
            ADD  IE
            SACL IA                       * IA = IA + IE
            BANZ JLOOP
  LAC  IE,2
  SACL IE                                 * IE = IE * 4
  B    KLOOP
*
* Digit reverse counter for radix-4 FFT computation.
*
DRC4     ZAC
         SACL J
         SACL I
         LARP 0
         LAR  AR0,HOLDN
         MAR  *-
         MAR  *-
DRLOOP          SUB  J
                BGEZ NOSWAP
* Swap I and J values.
                        OUT  I,PA0
                        IN   XI,PA1
                        IN   YI,PA1
                        OUT  J,PA0
                        IN   XJ,PA1
                        IN   YJ,PA1
                        OUT  J,PA0
                        OUT  XI,PA1
                        OUT  YI,PA1
                        OUT  I,PA0
                        OUT  XJ,PA1
                        OUT  YJ,PA1
*
NOSWAP          LAC  QUARTN,1
                SACL K
INLOOP                  LT   K
                        MPYK 3
                        PAC
                        SACL TEMP
```

```
                        SUB  J
                        BGZ  OUTL
                                LAC  J
                                SUB  TEMP
                                SACL J
                                LAC  K,14
                                SACH K
                                B    INLOOP
OUTL            LAC  J
                ADD  K
                SACL J
                LAC  I
                ADD  ONE,1
                SACL I
                BANZ DRLOOP
*
* End of FFT
*
STOP    B STOP
*
* Twiddle factor for theta = pi/4.
*
COS21 DATA 23169
*
* Sine and cosine table.
*
SINE  EQU $
      DATA 0
      DATA 3211
      DATA 6392
      DATA 9511
      DATA 12539
      DATA 15446
      DATA 18204
      DATA 20787
      DATA 23169
      DATA 25329
      DATA 27244
      DATA 28897
      DATA 30272
      DATA 31356
      DATA 32137
      DATA 32609
COSINE EQU $
      DATA 32767
      DATA 32609
      DATA 32137
      DATA 31356
      DATA 30272
      DATA 28897
      DATA 27244
      DATA 25329
      DATA 23169
      DATA 20787
      DATA 18204
      DATA 15446
      DATA 12539
```

```
DATA 9511
DATA 6392
DATA 3211
DATA 0
DATA -3211
DATA -6392
DATA -9511
DATA -12539
DATA -15446
DATA -18204
DATA -20787
DATA -23169
DATA -25329
DATA -27244
DATA -28897
DATA -30272
DATA -31356
DATA -32137
DATA -32609
DATA -32767
DATA -32609
DATA -32137
DATA -31356
DATA -30272
DATA -28897
DATA -27244
DATA -25329
DATA -23169
DATA -20787
DATA -18204
DATA -15446
DATA -12539
DATA -9511
DATA -6392
DATA -3211
END
```

```
        IDT 'FFTS64'
*
*
*    Cooley-Tukey Radix-4, DIF FFT Program - 64-point straight-line.
*
*       Three radix-4 butterflies - implemented with macros.
*       Complex input data located on page 0 of data memory.
*       Results are left in data RAM.
*       Uses 13-bit coefficients from MPYK instructions.
*       No scaling is done on intermediate values in the program.
*
*
* Data memory allocation.
*
X0       EQU 0            * X's are real data values
Y0       EQU 1            * Y's are imaginary data values
X1       EQU 2
Y1       EQU 3
X2       EQU 4
Y2       EQU 5
X3       EQU 6
Y3       EQU 7
X4       EQU 8
Y4       EQU 9
X5       EQU 10
Y5       EQU 11
X6       EQU 12
Y6       EQU 13
X7       EQU 14
Y7       EQU 15
X8       EQU 16
Y8       EQU 17
X9       EQU 18
Y9       EQU 19
X10      EQU 20
Y10      EQU 21
X11      EQU 22
Y11      EQU 23
X12      EQU 24
Y12      EQU 25
X13      EQU 26
Y13      EQU 27
X14      EQU 28
Y14      EQU 29
X15      EQU 30
Y15      EQU 31
X16      EQU 32
Y16      EQU 33
X17      EQU 34
Y17      EQU 35
X18      EQU 36
Y18      EQU 37
X19      EQU 38
Y19      EQU 39
X20      EQU 40
Y20      EQU 41
X21      EQU 42
```

```
Y21        EQU 43
X22        EQU 44
Y22        EQU 45
X23        EQU 46
Y23        EQU 47
X24        EQU 48
Y24        EQU 49
X25        EQU 50
Y25        EQU 51
X26        EQU 52
Y26        EQU 53
X27        EQU 54
Y27        EQU 55
X28        EQU 56
Y28        EQU 57
X29        EQU 58
Y29        EQU 59
X30        EQU 60
Y30        EQU 61
X31        EQU 62
Y31        EQU 63
X32        EQU 64
Y32        EQU 65
X33        EQU 66
Y33        EQU 67
X34        EQU 68
Y34        EQU 69
X35        EQU 70
Y35        EQU 71
X36        EQU 72
Y36        EQU 73
X37        EQU 74
Y37        EQU 75
X38        EQU 76
Y38        EQU 77
X39        EQU 78
Y39        EQU 79
X40        EQU 80
Y40        EQU 81
X41        EQU 82
Y41        EQU 83
X42        EQU 84
Y42        EQU 85
X43        EQU 86
Y43        EQU 87
X44        EQU 88
Y44        EQU 89
X45        EQU 90
Y45        EQU 91
X46        EQU 92
Y46        EQU 93
X47        EQU 94
Y47        EQU 95
X48        EQU 96
Y48        EQU 97
X49        EQU 98
Y49        EQU 99
```

```
X50       EQU 100
Y50       EQU 101
X51       EQU 102
Y51       EQU 103
X52       EQU 104
Y52       EQU 105
X53       EQU 106
Y53       EQU 107
X54       EQU 108
Y54       EQU 109
X55       EQU 110
Y55       EQU 111
X56       EQU 112
Y56       EQU 113
X57       EQU 114
Y57       EQU 115
X58       EQU 116
Y58       EQU 117
X59       EQU 118
Y59       EQU 119
X60       EQU 120
Y60       EQU 121
X61       EQU 122
Y61       EQU 123
X62       EQU 124
Y62       EQU 125
X63       EQU 126
Y63       EQU 127
T1        EQU 128        * Temporary locations addressed
T2        EQU 129        * by auxiliary registers.
*************************************************************************
*
* Cosine and sine values (13-bit) for coefficients.
*
CO0     EQU 4095
CO1     EQU 4076
CO2     EQU 4017
CO3     EQU 3919
CO4     EQU 3784
CO5     EQU 3612
CO6     EQU 3405
CO7     EQU 3166
CO8     EQU 2896
CO9     EQU 2598
CO10    EQU 2275
CO11    EQU 1930
CO12    EQU 1567
CO13    EQU 1188
CO14    EQU 799
CO15    EQU 401
CO16    EQU 0
CO17    EQU -401
CO18    EQU -799
CO19    EQU -1188
CO20    EQU -1567
CO21    EQU -1930
CO22    EQU -2275
```

```
CO23    EQU -2598
CO24    EQU -2896
CO25    EQU -3166
CO26    EQU -3405
CO27    EQU -3612
CO28    EQU -3784
CO29    EQU -3919
CO30    EQU -4017
CO31    EQU -4076
CO32    EQU -4095
CO33    EQU -4076
CO34    EQU -4017
CO35    EQU -3919
CO36    EQU -3784
CO37    EQU -3612
CO38    EQU -3405
CO39    EQU -3166
CO40    EQU -2896
CO41    EQU -2598
CO42    EQU -2275
CO43    EQU -1930
CO44    EQU -1567
CO45    EQU -1188
CO46    EQU -799
CO47    EQU -401
CO48    EQU 0
CO49    EQU 401
CO50    EQU 799
CO51    EQU 1188
CO52    EQU 1567
CO53    EQU 1930
CO54    EQU 2275
CO55    EQU 2598
CO56    EQU 2896
CO57    EQU 3166
CO58    EQU 3405
CO59    EQU 3612
CO60    EQU 3784
CO61    EQU 3919
CO62    EQU 4017
CO63    EQU 4076
*
SI0     EQU 0
SI1     EQU 401
SI2     EQU 799
SI3     EQU 1188
SI4     EQU 1567
SI5     EQU 1930
SI6     EQU 2275
SI7     EQU 2598
SI8     EQU 2896
SI9     EQU 3166
SI10    EQU 3405
SI11    EQU 3612
SI12    EQU 3784
SI13    EQU 3919
SI14    EQU 4017
```

```
SI15    EQU 4076
SI16    EQU 4095
SI17    EQU 4076
SI18    EQU 4017
SI19    EQU 3919
SI20    EQU 3784
SI21    EQU 3612
SI22    EQU 3405
SI23    EQU 3166
SI24    EQU 2896
SI25    EQU 2598
SI26    EQU 2275
SI27    EQU 1930
SI28    EQU 1567
SI29    EQU 1188
SI30    EQU 799
SI31    EQU 401
SI32    EQU 0
SI33    EQU -401
SI34    EQU -799
SI35    EQU -1188
SI36    EQU -1567
SI37    EQU -1930
SI38    EQU -2275
SI39    EQU -2598
SI40    EQU -2896
SI41    EQU -3166
SI42    EQU -3405
SI43    EQU -3612
SI44    EQU -3784
SI45    EQU -3919
SI46    EQU -4017
SI47    EQU -4076
SI48    EQU -4095
SI49    EQU -4076
SI50    EQU -4017
SI51    EQU -3919
SI52    EQU -3784
SI53    EQU -3612
SI54    EQU -3405
SI55    EQU -3166
SI56    EQU -2896
SI57    EQU -2598
SI58    EQU -2275
SI59    EQU -1930
SI60    EQU -1567
SI61    EQU -1188
SI62    EQU -799
SI63    EQU -401
************************************************************************
*
* MACROS TO PRODUCE STRAIGHT-LINE 64-POINT COMPLEX FFT.
*
************************************************************************
*
*  ZERO for case theta = 0.
*
```

```
*  X's and Y's are input and output locations for butterfly.
*
ZERO    $MACRO  XI,XI1,XI2,XI3,YI,YI1,YI2,YI3
*
        LAC  :XI1:
        ADD  :XI3:
        SACL :XI1:                * XI1 = XI1 + XI3
        SUB  :XI3:,1
        SACL :XI3:                * XI3 = XI1 - XI3
        LAC  :XI:
        SUB  :XI2:
        SACL *,0,AR1              * T1 = XI - XI2
        ADD  :XI2:,1              * R1 (ACC) = XI + XI2
        ADD  :XI1:
        SACL :XI:                 * XI = R1 + XI1
        SUB  :XI1:,1
        SACL :XI2:                * XI2 = R1 - XI1
*
        LAC  :YI1:
        ADD  :YI3:
        SACL :YI1:                * YI1 = YI1 + YI3
        SUB  :YI3:,1
        SACL :YI3:                * YI3 = YI1 - YI3
        LAC  :YI:
        SUB  :YI2:
        SACL *,0,AR0              * T2 = YI - YI2
        ADD  :YI2:,1              * R1 (ACC) = YI + YI2
        ADD  :YI1:
        SACL :YI:                 * YI = R1 + YI1
        SUB  :YI1:,1
        SACL :YI2:                * YI2 = R1 - YI1
*
        LAC  *
        ADD  :YI3:
        SACL :XI1:                * XI1 = T1 + YI3
        SUB  :YI3:,1
        SACL *,0,AR1              * T1 = T1 - YI3
        LAC  *,0,AR0
        ADD  :XI3:
        SACL :YI3:                * YI3 = T2 + XI3
        SUB  :XI3:,1
        SACL :YI1:                * YI1 = T2 - XI3
        LAC  *
        SACL :XI3:                * XI3 = T1
*
        $END
***************************************************************************
*
* NORM - standard radix-4 butterfly.
*
* X's and Y's are locations of data input and output.
* IA's specify twiddle factor locations.
*
NORM    $MACRO  XI,XI1,XI2,XI3,YI,YI1,YI2,YI3,IA1,IA2,IA3
*
        LAC  :XI1:
        ADD  :XI3:
```

```
        SACL :XI1:                  * XI1 = XI1 + XI3
        SUB  :XI3:,1
        SACL :XI3:                  * XI3 = XI1 - XI3
        LAC  :XI:
        SUB  :XI2:
        SACL *,0,AR1                * T1 = XI - XI2
        ADD  :XI2:,1                * R1 (ACC) = XI + XI2
        ADD  :XI1:
        SACL :XI:                   * XI = R1 + XI1
        SUB  :XI1:,1
        SACL :XI2:                  * XI2 = R1 - XI1
*
        LAC  :YI1:
        ADD  :YI3:
        SACL :YI1:                  * YI1 = YI1 + YI3
        SUB  :YI3:,1
        SACL :YI3:                  * YI3 = YI1 - YI3
        LAC  :YI:
        SUB  :YI2:
        SACL *,0,AR0                * T2 = YI - YI2
        ADD  :YI2:,1                * R1 (ACC) = YI + YI2
        ADD  :YI1:
        SACL :YI:                   * YI = R1 + YI1
        SUB  :YI1:,1
        SACL :YI2:                  * YI2 = R1 - YI1
*
        LAC  *
        ADD  :YI3:
        SACL :XI1:                  * XI1 = T1 + YI3
        SUB  :YI3:,1
        SACL *,0,AR1                * T1 = T1 - YI3
        LAC  *
        ADD  :XI3:
        SACL *                      * T2 = T2 + XI3
        SUB  :XI3:,1
        SACL :YI1:                  * YI1 = T2 - XI3
*
        LT   :YI1:
        MPYK CO:IA1:
        PAC
        LT   :XI1:
        MPYK SI:IA1:
        SPAC
        LT   :YI1:
        SACH :YI1:,4                * YI1 = YI1*CO1 - XI1*SI1
        MPYK SI:IA1:
        PAC
        LT   :XI1:
        MPYK CO:IA1:
        LTA  :YI2:
        SACH :XI1:,4                * XI1 = YI1*SI1 + XI1*CO1
*
        MPYK CO:IA2:
        PAC
        LT   :XI2:
        MPYK SI:IA2:
        SPAC
```

```
        LT   :YI2:
        SACH :YI2:,4              * YI2 = YI2*CO2 - XI2*SI2
        MPYK SI:IA2:
        PAC
        LT   :XI2:
        MPYK CO:IA2:
        LTA  *,AR0
        SACH :XI2:,4              * XI2 = YI2*SI2 + XI2*CO2
*
        MPYK CO:IA3:
        PAC
        LT   *,AR1
        MPYK SI:IA3:
        SPAC
        SACH :YI3:,4              * YI3 = T2*CO3 - T1*SI3
        MPYK CO:IA3:
        PAC
        LT   *,AR0
        MPYK SI:IA3:
        APAC
        SACH :XI3:,4              * XI3 = T1*CO3 + T2*SI3
*
        $END
*************************************************************************
*
* SPEC for case theta = pi/4
*
* X's and Y's are data input and output locations.
*
SPEC    $MACRO  XI,XI1,XI2,XI3,YI,YI1,YI2,YI3
*
        LAC  :XI:
        ADD  :XI2:
        SACL :XI:                 * XI = XI + XI2
        SUB  :XI2:,1
        SACL :XI2:                * XI2 = XI - XI2
*
        LAC  :YI:
        ADD  :YI2:
        SACL :YI:                 * YI = YI + YI2
        SUB  :YI2:,1
        SACL :YI2:                * YI2 = YI - YI2
*
        LAC  :XI1:
        ADD  :XI3:
        SACL :XI1:                * XI1 = XI1 + XI3
        SUB  :XI3:,1
        SACL :XI3:                * XI3 = XI1 - XI3
*
        LAC  :YI1:
        ADD  :YI3:
        SACL :YI1:                * YI1 = YI1 + YI3
        SUB  :YI3:,1
        SACL :YI3:                * YI3 = YI1 - YI3
*
        LAC  :XI1:
        SUB  :XI:
```

```
        SACL *,0,AR1              * T1 = XI1 - XI
        ADD  :XI:,1
        SACL :XI:                 * XI = XI1 + XI
        LAC  :YI:
        ADD  :YI1:
        SACL :YI:                 * YI = YI + YI1
        SUB  :YI1:,1
        SACL *                    * T2 = YI - YI1
*
        LAC  :XI2:
        ADD  :YI3:
        SACL :XI1:                * XI1 = XI2 + YI3
        SUB  :YI3:,1
        SACL :YI3:                * YI3 = XI2 - YI3
        LAC  *,0,AR0
        SACL :XI2:                * XI2 = T2
        LAC  :YI2:
        SUB  :XI3:
        SACL :YI1:                * YI1 = YI2 - XI3
        ADD  :XI3:,1
        SACL :XI3:                * XI3 = YI2 + XI3
        LAC  *
        SACL :YI2:                * YI2 = T1
*
        LT   :YI1:
        MPYK CO8
        PAC
        LT   :XI1:
        MPYK CO8
        SPAC
        SACH :YI1:,4              * YI1 = (YI1-XI1)*CO8
        APAC
        LTA  :YI3:
        SACH :XI1:,4              * XI1 = (YI1+XI1)*CO8
        ZAC
        MPYK CO8
        SPAC
        LT   :XI3:
        MPYK CO8
        SPAC
        SACH :YI3:,4              * YI3 = -(YI3+XI3)*CO8
        APAC
        APAC
        SACH :XI3:,4              * XI3 = (XI3 - YI3)*CO8
*
        $END
************************************************************************
*
*       MAIN routine to call above macros with appropriate parameters.
*
************************************************************************
        AORG 0
*
FFT64 LDPK 0                      * Use input data is on page 0.
      LARP 0
      LARK AR0,T1                 * AR0 points to Temp 1.
      LARK AR1,T2                 * AR1 points to Temp 2.
```

```
*
* PASS 1
*
      ZERO    X0,X16,X32,X48,Y0,Y16,Y32,Y48
*
      NORM    X1,X17,X33,X49,Y1,Y17,Y33,Y49,1,2,3
      NORM    X2,X18,X34,X50,Y2,Y18,Y34,Y50,2,4,6
      NORM    X3,X19,X35,X51,Y3,Y19,Y35,Y51,3,6,9
      NORM    X4,X20,X36,X52,Y4,Y20,Y36,Y52,4,8,12
      NORM    X5,X21,X37,X53,Y5,Y21,Y37,Y53,5,10,15
      NORM    X6,X22,X38,X54,Y6,Y22,Y38,Y54,6,12,18
      NORM    X7,X23,X39,X55,Y7,Y23,Y39,Y55,7,14,21
*
      SPEC    X8,X24,X40,X56,Y8,Y24,Y40,Y56
*
      NORM    X9,X25,X41,X57,Y9,Y25,Y41,Y57,9,18,27
      NORM    X10,X26,X42,X58,Y10,Y26,Y42,Y58,10,20,30
      NORM    X11,X27,X43,X59,Y11,Y27,Y43,Y59,11,22,33
      NORM    X12,X28,X44,X60,Y12,Y28,Y44,Y60,12,24,36
      NORM    X13,X29,X45,X61,Y13,Y29,Y45,Y61,13,26,39
      NORM    X14,X30,X46,X62,Y14,Y30,Y46,Y62,14,28,42
      NORM    X15,X31,X47,X63,Y15,Y31,Y47,Y63,15,30,45
*
* PASS 2
*
      ZERO    X0,X4,X8,X12,Y0,Y4,Y8,Y12
      ZERO    X16,X20,X24,X28,Y16,Y20,Y24,Y28
      ZERO    X32,X36,X40,X44,Y32,Y36,Y40,Y44
      ZERO    X48,X52,X56,X60,Y48,Y52,Y56,Y60
*
      NORM    X1,X5,X9,X13,Y1,Y5,Y9,Y13,4,8,12
      NORM    X17,X21,X25,X29,Y17,Y21,Y25,Y29,4,8,12
      NORM    X33,X37,X41,X45,Y33,Y37,Y41,Y45,4,8,12
      NORM    X49,X53,X57,X61,Y49,Y53,Y57,Y61,4,8,12
*
      SPEC    X2,X6,X10,X14,Y2,Y6,Y10,Y14
      SPEC    X18,X22,X26,X30,Y18,Y22,Y26,Y30
      SPEC    X34,X38,X42,X46,Y34,Y38,Y42,Y46
      SPEC    X50,X54,X58,X62,Y50,Y54,Y58,Y62
*
      NORM    X3,X7,X11,X15,Y3,Y7,Y11,Y15,12,24,36
      NORM    X19,X23,X27,X31,Y19,Y23,Y27,Y31,12,24,36
      NORM    X35,X39,X43,X47,Y35,Y39,Y43,Y47,12,24,36
      NORM    X51,X55,X59,X63,Y51,Y55,Y59,Y63,12,24,36
*
*  PASS 3
*
      ZERO    X0,X1,X2,X3,Y0,Y1,Y2,Y3
      ZERO    X4,X5,X6,X7,Y4,Y5,Y6,Y7
      ZERO    X8,X9,X10,X11,Y8,Y9,Y10,Y11
      ZERO    X12,X13,X14,X15,Y12,Y13,Y14,Y15
      ZERO    X16,X17,X18,X19,Y16,Y17,Y18,Y19
      ZERO    X20,X21,X22,X23,Y20,Y21,Y22,Y23
      ZERO    X24,X25,X26,X27,Y24,Y25,Y26,Y27
      ZERO    X28,X29,X30,X31,Y28,Y29,Y30,Y31
      ZERO    X32,X33,X34,X35,Y32,Y33,Y34,Y35
      ZERO    X36,X37,X38,X39,Y36,Y37,Y38,Y39
```

```
        ZERO     X40,X41,X42,X43,Y40,Y41,Y42,Y43
        ZERO     X44,X45,X46,X47,Y44,Y45,Y46,Y47
        ZERO     X48,X49,X50,X51,Y48,Y49,Y50,Y51
        ZERO     X52,X53,X54,X55,Y52,Y53,Y54,Y55
        ZERO     X56,X57,X58,X59,Y56,Y57,Y58,Y59
        ZERO     X60,X61,X62,X63,Y60,Y61,Y62,Y63
*
        END
```

```
        IDT 'FFT8'
*
*     Cooley-Tukey Radix-8, DIF FFT Program
*
*        Two radix-8 butterflies.
*        Complex input data - size limited only by program memory availability.
*        Uses table lookup of twiddle factors.
*        No scaling is done on intermediate values in the program.
*        External data RAM is addressed via peripheral I/O instructions.
*                An address counter is required and is loaded by a write
*                to port 0.  Data is read from and written to port 1.
*        Data in the external RAM assumes complex data with corresponding
*                real and imaginary data values in consecutive locations.
*
* N is the size of the transform (N = 8**M).
N EQU 64
M EQU 2
*
* Data memory allocation.
*
XI1     EQU 0               ** Real part of data
XI2     EQU 1               *
XI3     EQU 2               *
XI4     EQU 3               *
XI5     EQU 4               *
XI6     EQU 5               *
XI7     EQU 6               *
XI8     EQU 7               **
YI1     EQU 8               ** Imaginary part of data
YI2     EQU 9               *
YI3     EQU 10              *
YI4     EQU 11              *
YI5     EQU 12              *
YI6     EQU 13              *
YI7     EQU 14              *
YI8     EQU 15              **
CO2     EQU 16              ** Real part of twiddle factors
CO3     EQU 17              *
CO4     EQU 18              *
CO5     EQU 19              *
CO6     EQU 20              *
CO7     EQU 21              *
CO8     EQU 22              **
SI2     EQU 23              ** Imaginary part of twiddle factors
SI3     EQU 24              *
SI4     EQU 25              *
SI5     EQU 26              *
SI6     EQU 27              *
SI7     EQU 28              *
SI8     EQU 29              **
C81     EQU 30              ** Twiddle factor for theta = pi/4
R1      EQU 31              ** Temporary data locations
R2      EQU 32              *
R3      EQU 33              *
R4      EQU 34              *
R5      EQU 35              *
R6      EQU 36              *
```

```
R7      EQU 37            *
R8      EQU 38            *
S1      EQU 39            *
S2      EQU 40            *
S3      EQU 41            *
S4      EQU 42            *
S5      EQU 43            *
S6      EQU 44            *
S7      EQU 45            *
S8      EQU 46            *
T1      EQU 47            *
T2      EQU 48            *
XI      EQU 49            *
YI      EQU 50            *
XJ      EQU 51            *
YJ      EQU 52            *
TEMP    EQU 53            **
I1      EQU 54            ** External addresses of data
I2      EQU 55            *
I3      EQU 56            *
I4      EQU 57            *
I5      EQU 58            *
I6      EQU 59            *
I7      EQU 60            *
I8      EQU 61            **
IA      EQU 62            * Twiddle factor index
IE      EQU 63            * Increment of IA
HOLDN   EQU 64            * Contains value N
QUARTN  EQU 65            * Contains value N/4
PIE4    EQU 66            * Contains value N/8
CTABLE  EQU 67            * Start of real twiddle factor table
STABLE  EQU 68            * Start of imaginary twiddle factor table
ONE     EQU 69            * Contains value 1
I       EQU 70            ** Counters
J       EQU 71            *
K       EQU 72            *
N2      EQU 73            *
N1      EQU 74            **
*
* Start of program memory section.
*
          AORG 0
START     LDPK 0
          LACK 1
          SACL IE                 * Initialize IE = 1
          SACL ONE                * ONE = 1
          LT   ONE
          MPYK N
          PAC
          SACL HOLDN
          SACL N2                 * Initialize N2 = N
          LAC  HOLDN,13
          SACH QUARTN,1           * QUARTN = N/4
          SACH PIE4               * PIE4 = N/8
          MPYK SINE
          PAC
          SACL STABLE             * STABLE contains address of sine table
```

```
         ADD  QUARTN
         SACL CTABLE          * CTABLE contains address of cosine table
         MPYK COS81
         PAC
         TBLR C81             * Read cos(pi/4)
         LARK AR1,M-1         * AR0 contains K value
         LARP 0
***********************************************************************
*
* Butterfly for case theta = 0.
*
***********************************************************************
KLOOP    LAC  N2
         SACL N1              * N1 = N2
         LAC  N2,13
         SACH N2              * N2 = N2/8
         ZAC
         SACL I1              * I1 = 0
         LAR  AR0,IE
         MAR  *-
         LT   C81             * Coefficient is always C81 for this
*                             * butterfly.
*
* Start calculation of butterfly.
*
ZERO        LAC  I1
            ADD  N2,1
            SACL I2           * I2 = I1 + N2
            ADD  N2,1
            SACL I3           * I3 = I2 + N2
            ADD  N2,1
            SACL I4           * I4 = I3 + N2
            ADD  N2,1
            SACL I5           * I5 = I4 + N2
            ADD  N2,1
            SACL I6           * I6 = I5 + N2
            ADD  N2,1
            SACL I7           * I7 = I6 + N2
            ADD  N2,1
            SACL I8           * I8 = I7 + N2
*
* Read in eight data values.
*
            OUT  I1,PA0
            IN   XI1,PA1
            IN   YI1,PA1
            OUT  I2,PA0
            IN   XI2,PA1
            IN   YI2,PA1
            OUT  I3,PA0
            IN   XI3,PA1
            IN   YI3,PA1
            OUT  I4,PA0
            IN   XI4,PA1
            IN   YI4,PA1
            OUT  I5,PA0
            IN   XI5,PA1
```

```
          IN   YI5,PA1
          OUT  I6,PA0
          IN   XI6,PA1
          IN   YI6,PA1
          OUT  I7,PA0
          IN   XI7,PA1
          IN   YI7,PA1
          OUT  I8,PA0
          IN   XI8,PA1
          IN   YI8,PA1
*
* Main calculation.
*
          LAC  XI1
          ADD  XI5
          SACL R1                * R1 = X(I1) + X(I5)
          SUB  XI5,1
          SACL R5                * R5 = X(I1) - X(I5)
          LAC  XI2
          ADD  XI6
          SACL R2                * R2 = X(I2) + X(I6)
          SUB  XI6,1
          SACL R6                * R6 = X(I2) - X(I6)
          LAC  XI3
          ADD  XI7
          SACL R3                * R3 = X(I3) + X(I7)
          SUB  XI7,1
          SACL R7                * R7 = X(I3) - X(I7)
          LAC  XI4
          ADD  XI8
          SACL R4                * R4 = X(I4) + X(I8)
          SUB  XI8,1
          SACL R8                * R8 = X(I4) - X(I8)
          LAC  R1
          SUB  R3
          SACL T1                * T1 = R1 - R3
          ADD  R3,1
          SACL R1                * R1 = R1 + R3
          LAC  R2
          SUB  R4
          SACL R3                * R3 = R2 - R4
          ADD  R4,1
          SACL R2                * R2 = R2 + R4
          ADD  R1
          SACL XI1               * X(I1) = R1 + R2
          SUB  R2,1
          SACL XI5               * X(I5) = R1 - R2
          LAC  YI1
          ADD  YI5
          SACL R1                * R1 = Y(I1) + Y(I5)
          SUB  YI5,1
          SACL S5                * S5 = Y(I1) - Y(I5)
          LAC  YI2
          ADD  YI6
          SACL R2                * R2 = Y(I2) + Y(I6)
          SUB  YI6,1
          SACL S6                * S6 = Y(I2) - Y(I6)
```

```
        LAC  YI3
        ADD  YI7
        SACL S3              * S3 = Y(I3) + Y(I7)
        SUB  YI7,1
        SACL S7              * S7 = Y(I3) - Y(I7)
        LAC  YI4
        ADD  YI8
        SACL R4              * R4 = Y(I4) + Y(I8)
        SUB  YI8,1
        SACL S8              * S8 = Y(I4) - Y(I8)
        LAC  R1
        SUB  S3
        SACL T2              * T2 = R1 - S3
        ADD  S3,1
        SACL R1              * R1 = R1 + S3
        LAC  R2
        SUB  R4
        SACL S3              * S3 = R2 - R4
        ADD  R4,1
        SACL R2              * R2 = R2 + R4
        ADD  R1
        SACL YI1             * Y(I1) = R1 + R2
        SUB  R2,1
        SACL YI5             * Y(I5) = R1 - R2
        LAC  T1
        ADD  S3
        SACL XI3             * X(I3) = T1 + S3
        SUB  S3,1
        SACL XI7             * X(I7) = T1 - S3
        LAC  T2
        SUB  R3
        SACL YI3             * Y(I3) = T2 - R3
        ADD  R3,1
        SACL YI7             * Y(I7) = T2 + R3
*
        MPY  R6
        PAC
        MPY  R8
        SPAC
        SACH R1,1            * R1 = (R6 - R8) * C81
        APAC
        APAC
        SACH R6,1            * R6 = (R6 + R8) * C81
        MPY  S6
        PAC
        MPY  S8
        SPAC
        SACH R2,1            * R2 = (S6 - S8) * C81
        APAC
        APAC
        SACH S6,1            * S6 = (S6 + S8) * C81
*
        LAC  R5
        SUB  R1
        SACL T1              * T1 = R5 - R1
        ADD  R1,1
        SACL R5              * R5 = R5 + R1
```

```
            LAC  R7
            SUB  R6
            SACL R8             * R8 = R7 - R6
            ADD  R6,1
            SACL R7             * R7 = R7 + R6
            LAC  S5
            SUB  R2
            SACL T2             * T2 = S5 - R2
            ADD  R2,1
            SACL S5             * S5 = S5 + R2
            LAC  S7
            SUB  S6
            SACL S8             * S8 = S7 - S6
            ADD  S6,1
            SACL S7             * S7 = S7 + S6
            ADD  R5
            SACL XI2            * X(I2) = R5 + S7
            SUB  S7,1
            SACL XI8            * X(I8) = R5 - S7
            LAC  T1
            ADD  S8
            SACL XI6            * X(I6) = T1 + S8
            SUB  S8,1
            SACL XI4            * X(I4) = T1 - S8
            LAC  S5
            SUB  R7
            SACL YI2            * Y(I2) = S5 - R7
            ADD  R7,1
            SACL YI8            * Y(I8) = S5 + R7
            LAC  T2
            SUB  R8
            SACL YI6            * Y(I6) = T2 - R8
            ADD  R8,1
            SACL YI4            * Y(I4) = T2 + R8
*
* Output eight data values.
*
            OUT  I1,PA0
            OUT  XI1,PA1
            OUT  YI1,PA1
            OUT  I2,PA0
            OUT  XI2,PA1
            OUT  YI2,PA1
            OUT  I3,PA0
            OUT  XI3,PA1
            OUT  YI3,PA1
            OUT  I4,PA0
            OUT  XI4,PA1
            OUT  YI4,PA1
            OUT  I5,PA0
            OUT  XI5,PA1
            OUT  YI5,PA1
            OUT  I6,PA0
            OUT  XI6,PA1
            OUT  YI6,PA1
            OUT  I7,PA0
            OUT  XI7,PA1
```

```
            OUT  YI7,PA1
            OUT  I8,PA0
            OUT  XI8,PA1
            OUT  YI8,PA1
*
            LAC  I1
            ADD  N1,1
            SACL I1                  * I1 = I1 + N1
            BANZ ZERO                * Loop while I1 < N
*
        LARP 1
        BANZ ISOK                    * After last phase, go to end of program.
        B    DRC8
*******************************************************************************
*
* Standard radix-8 butterfly.
*
*******************************************************************************
ISOK    ZAC
        SACL IA                      * IA = 0
        LAR  AR0,N2
        LARP 0
        MAR  *-
        MAR  *-                      * Loop for J = 1 to N2-1
        LAC  ONE
        SACL J                       * J = 1
JLOOP       LAC  IA
            ADD  IE
            SACL IA                  * IA = IA + IE
*
* Table lookup of twiddle factors.
*
                LAC  CTABLE          * Get cosines
                ADD  IA
                TBLR CO2
                ADD  IA
                TBLR CO3
                ADD  IA
                TBLR CO4
                ADD  IA
                TBLR CO5
                ADD  IA
                TBLR CO6
                ADD  IA
                TBLR CO7
                ADD  IA
                TBLR CO8
                LAC  STABLE          * Get sines
                ADD  IA
                TBLR SI2
                ADD  IA
                TBLR SI3
                ADD  IA
                TBLR SI4
                ADD  IA
                TBLR SI5
                ADD  IA
```

```
              TBLR SI6
              ADD  IA
              TBLR SI7
              ADD  IA
              TBLR SI8
*
* Start butterfly calculation.
*
              LAC  J,1
              SACL I1                * I1 = J (data organized as real value followed
*                                    * by imaginary so address I1 is 2 times J).
*
NORM       LAC  I1
           ADD  N2,1
           SACL I2                   * I2 = I1 + N2
           ADD  N2,1
           SACL I3                   * I3 = I2 + N2
           ADD  N2,1
           SACL I4                   * I4 = I3 + N2
           ADD  N2,1
           SACL I5                   * I5 = I4 + N2
           ADD  N2,1
           SACL I6                   * I6 = I5 + N2
           ADD  N2,1
           SACL I7                   * I7 = I6 + N2
           ADD  N2,1
           SACL I8                   * I8 = I7 + N2
*
* Input eight data values.
*
           OUT  I1,PA0
           IN   XI1,PA1
           IN   YI1,PA1
           OUT  I2,PA0
           IN   XI2,PA1
           IN   YI2,PA1
           OUT  I3,PA0
           IN   XI3,PA1
           IN   YI3,PA1
           OUT  I4,PA0
           IN   XI4,PA1
           IN   YI4,PA1
           OUT  I5,PA0
           IN   XI5,PA1
           IN   YI5,PA1
           OUT  I6,PA0
           IN   XI6,PA1
           IN   YI6,PA1
           OUT  I7,PA0
           IN   XI7,PA1
           IN   YI7,PA1
           OUT  I8,PA0
           IN   XI8,PA1
           IN   YI8,PA1
*
* Main calculation in butterfly
*
```

```
        LAC   XI1
        ADD   XI5
        SACL  R1                * R1 = X(I1) + X(I5)
        SUB   XI5,1
        SACL  R5                * R5 = X(I1) - X(I5)
        LAC   XI2
        ADD   XI6
        SACL  R2                * R2 = X(I2) + X(I6)
        SUB   XI6,1
        SACL  R6                * R6 = X(I2) - X(I6)
        LAC   XI3
        ADD   XI7
        SACL  R3                * R3 = X(I3) + X(I7)
        SUB   XI7,1
        SACL  R7                * R7 = X(I3) - X(I7)
        LAC   XI4
        ADD   XI8
        SACL  R4                * R4 = X(I4) + X(I8)
        SUB   XI8,1
        SACL  R8                * R8 = X(I4) - X(I8)
        LAC   R1
        SUB   R3
        SACL  T1                * T1 = R1 - R3
        ADD   R3,1
        SACL  R1                * R1 = R1 + R3
        LAC   R2
        SUB   R4
        SACL  R3                * R3 = R2 - R4
        ADD   R4,1
        SACL  R2                * R2 = R2 + R4
        ADD   R1
        SACL  XI1               * X(I1) = R1 + R2
        SUB   R2,1
        SACL  R2                * R2 = R1 - R2
*
        LAC   YI1
        ADD   YI5
        SACL  S1                * S1 = Y(I1) + Y(I5)
        SUB   YI5,1
        SACL  S5                * S5 = Y(I1) - Y(I5)
        LAC   YI2
        ADD   YI6
        SACL  S2                * S2 = Y(I2) + Y(I6)
        SUB   YI6,1
        SACL  S6                * S6 = Y(I2) - Y(I6)
        LAC   YI3
        ADD   YI7
        SACL  S3                * S3 = Y(I3) + Y(I7)
        SUB   YI7,1
        SACL  S7                * S7 = Y(I3) - Y(I7)
        LAC   YI4
        ADD   YI8
        SACL  S4                * S4 = Y(I4) + Y(I8)
        SUB   YI8,1
        SACL  S8                * S8 = Y(I4) - Y(I8)
        LAC   S1
        SUB   S3
```

```
         SACL T2               * T2 = S1 - S3
         ADD  S3,1
         SACL S1               * S1 = S1 + S3
         LAC  S2
         SUB  S4
         SACL S3               * S3 = S2 - S4
         ADD  S4,1
         SACL S2               * S2 = S2 + S4
         ADD  S1
         SACL YI1              * Y(I1) = S1 + S2
         SUB  S2,1
         SACL S2               * S2 = S1 - S2
         LAC  T1
         ADD  S3
         SACL R1               * R1 = T1 + S3
         SUB  S3,1
         SACL T1               * T1 = T1 - S3
         LAC  T2
         SUB  R3
         SACL S1               * S1 = T2 - R3
         ADD  R3,1
         SACL T2               * T2 = T2 + R3
*
         LT   CO5
         MPY  S2
         PAC
         LT   SI5
         MPY  R2
         SPAC
         SACH YI5,1            * Y(I5) = CO5*S2 - SI5*R2
         MPY  S2
         PAC
         LT   CO5
         MPY  R2
         LTA  CO3
         SACH XI5,1            * X(I5) = CO5*R2 + SI5*S2
         MPY  S1
         PAC
         LT   SI3
         MPY  R1
         SPAC
         SACH YI3,1            * Y(I3) = CO3*S1 - SI3*R1
         MPY  S1
         PAC
         LT   CO3
         MPY  R1
         LTA  CO7
         SACH XI3,1            * X(I3) = CO3*R1 + SI3*S1
         MPY  T2
         PAC
         LT   SI7
         MPY  T1
         SPAC
         SACH YI7,1            * Y(I7) = CO7*T2 - SI7*T1
         MPY  T2
         PAC
         LT   CO7
```

```
            MPY  T1
            LTA  C81
            SACH XI7,1              * X(I7) = CO7*T1 + SI7*T2
*
            MPY  R6
            PAC
            MPY  R8
            SPAC
            SACH R1,1               * R1 = (R6 - R8) * C81
            APAC
            APAC
            SACH R6,1               * R6 = (R6 + R8) * C81
            MPY  S6
            PAC
            MPY  S8
            SPAC
            SACH S1,1               * S1 = (S6 - S8) * C81
            APAC
            LTA  CO2
            SACH S6,1               * S6 = (S6 + S8) * C81
*
            LAC  R5
            SUB  R1
            SACL T1                 * T1 = R5 - R1
            ADD  R1,1
            SACL R5                 * R5 = R5 + R1
            LAC  R7
            SUB  R6
            SACL R8                 * R8 = R7 - R6
            ADD  R6,1
            SACL R7                 * R7 = R7 + R6
            LAC  S5
            SUB  S1
            SACL T2                 * T2 = S5 - S1
            ADD  S1,1
            SACL S5                 * S5 = S5 + S1
            LAC  S7
            SUB  S6
            SACL S8                 * S8 = S7 - S6
            ADD  S6,1
            SACL S7                 * S7 = S7 + S6
            ADD  R5
            SACL R1                 * R1 = R5 + S7
            SUB  S7,1
            SACL R5                 * R5 = R5 - S7
            LAC  T1
            ADD  S8
            SACL R6                 * R6 = T1 + S8
            SUB  S8,1
            SACL T1                 * T1 = T1 - S8
            LAC  S5
            SUB  R7
            SACL S1                 * S1 = S5 - R7
            ADD  R7,1
            SACL S5                 * S5 = S5 + R7
            LAC  T2
            SUB  R8
```

```
          SACL S6          * S6 = T2 - R8
          ADD  R8,1
          SACL T2          * T2 = T2 + R8
*
          MPY  S1
          PAC
          LT   SI2
          MPY  R1
          SPAC
          SACH YI2,1       * Y(I2) = CO2*S1 - SI2*R1
          MPY  S1
          PAC
          LT   CO2
          MPY  R1
          LTA  CO8
          SACH XI2,1       * X(I2) = CO2*R1 + SI2*S1
          MPY  S5
          PAC
          LT   SI8
          MPY  R5
          SPAC
          SACH YI8,1       * Y(I8) = CO8*S5 - SI8*R5
          MPY  S5
          PAC
          LT   CO8
          MPY  R5
          LTA  CO6
          SACH XI8,1       * X(I8) = CO8*R5 + SI8*S5
          MPY  S6
          PAC
          LT   SI6
          MPY  R6
          SPAC
          SACH YI6,1       * Y(I6) = CO6*S6 - SI6*R6
          MPY  S6
          PAC
          LT   CO6
          MPY  R6
          LTA  CO4
          SACH XI6,1       * X(I6) = CO6*R6 + SI6*S6
          MPY  T2
          PAC
          LT   SI4
          MPY  T1
          SPAC
          SACH YI4,1       * Y(I4) = CO4*T2 - SI4*T1
          MPY  T2
          PAC
          LT   CO4
          MPY  T1
          APAC
          SACH XI4,1       * X(I4) = CO4*T1 + SI4*T2
*
* Output results.
*
          OUT  I1,PA0
          OUT  XI1,PA1
```

```
            OUT  YI1,PA1
            OUT  I2,PA0
            OUT  XI2,PA1
            OUT  YI2,PA1
            OUT  I3,PA0
            OUT  XI3,PA1
            OUT  YI3,PA1
            OUT  I4,PA0
            OUT  XI4,PA1
            OUT  YI4,PA1
            OUT  I5,PA0
            OUT  XI5,PA1
            OUT  YI5,PA1
            OUT  I6,PA0
            OUT  XI6,PA1
            OUT  YI6,PA1
            OUT  I7,PA0
            OUT  XI7,PA1
            OUT  YI7,PA1
            OUT  I8,PA0
            OUT  XI8,PA1
            OUT  YI8,PA1
*
                  LAC  I1
                  ADD  N1,1
                  SACL I1              * I1 = I1 + N1
                  SUB  HOLDN,1
                  BLZ NORM             * Loop while I1 < N
               LAC  J
               ADD  ONE
               SACL J                  * J = J + 1
               BANZ JLOOP
       LAC  IE,3
       SACL IE                         * IE = IE * 8
       B KLOOP
*
* Digit reverse counter for radix-8 FFT computation.
*
DRC8     ZAC
         SACL J
         SACL I
         LARP 0
         LAR  AR0,HOLDN
         MAR  *-
         MAR  *-
DRLOOP          SUB  J
                BGEZ NOSWAP
* Swap I and J values.
                      OUT  I,PA0
                      IN   XI,PA1
                      IN   YI,PA1
                      OUT  J,PA0
                      IN   XJ,PA1
                      IN   YJ,PA1
                      OUT  J,PA0
                      OUT  XI,PA1
                      OUT  YI,PA1
```

```
                              OUT   I,PA0
                              OUT   XJ,PA1
                              OUT   YJ,PA1
NOSWAP            LAC   QUARTN
                  SACL  K
INLOOP                        LT    K
                              MPYK  7
                              PAC
                              SACL  TEMP
                              SUB   J
                              BGZ   OUTL
                                       LAC   J
                                       SUB   TEMP
                                       SACL  J
                                       LAC   K,13
                                       SACH  K
                                       B     INLOOP
OUTL              LAC   J
                  ADD   K
                  SACL  J
                  LAC   I
                  ADD   ONE,1
                  SACL  I
                  BANZ  DRLOOP
*
* End of FFT.
*
STOP     B STOP
*
* Twiddle factor for theta = pi/4'
*
COS81 DATA 23169
*
* Sine and cosine tables.
*
SINE  EQU $
      DATA 0
      DATA 3211
      DATA 6392
      DATA 9511
      DATA 12539
      DATA 15446
      DATA 18204
      DATA 20787
      DATA 23169
      DATA 25329
      DATA 27244
      DATA 28897
      DATA 30272
      DATA 31356
      DATA 32137
      DATA 32609
COSINE EQU $
      DATA 32767
      DATA 32609
      DATA 32137
      DATA 31356
```

```
DATA 30272
DATA 28897
DATA 27244
DATA 25329
DATA 23169
DATA 20787
DATA 18204
DATA 15446
DATA 12539
DATA 9511
DATA 6392
DATA 3211
DATA 0
DATA -3211
DATA -6392
DATA -9511
DATA -12539
DATA -15446
DATA -18204
DATA -20787
DATA -23169
DATA -25329
DATA -27244
DATA -28897
DATA -30272
DATA -31356
DATA -32137
DATA -32609
DATA -32767
DATA -32609
DATA -32137
DATA -31356
DATA -30272
DATA -28897
DATA -27244
DATA -25329
DATA -23169
DATA -20787
DATA -18204
DATA -15446
DATA -12539
DATA -9511
DATA -6392
DATA -3211
DATA 0
DATA 3211
DATA 6392
DATA 9511
DATA 12539
DATA 15446
DATA 18204
DATA 20787
DATA 23169
DATA 25329
DATA 27244
DATA 28897
DATA 30272
DATA 31356
DATA 32137
DATA 32609
END
```

```
          IDT 'PFA'
*
*   Prime-Factor FFT Program
*
*        Complex input data.
*        Table lookup of coefficients.
*        No scaling is done on intermediate values in the program.
*        External data RAM is addressed via peripheral I/O instructions.
*                An address counter is required and is loaded by a write
*                to port 0.  Data is read from and written to port 1.
*                Final results are output in unscrambled order to port 2.
*        Data in the external RAM assumes complex data with corresponding
*                real and imaginary data values in consecutive locations.
*
* N is the size of the transform (N = NI1*NI2*...NIm).
N        EQU 63
NI1      EQU 7
NI2      EQU 9
*
* Data Memory Allocation.
*
X1       EQU 0               *** Current data points
Y1       EQU 1               *
X2       EQU 2               *
Y2       EQU 3               *
X3       EQU 4               *
Y3       EQU 5               *
X4       EQU 6               *
Y4       EQU 7               *
X5       EQU 8               *
Y5       EQU 9               *
X6       EQU 10              *
Y6       EQU 11              *
X7       EQU 12              *
Y7       EQU 13              *
X8       EQU 14              *
Y8       EQU 15              *
X9       EQU 16              *
Y9       EQU 17              ***
I1       EQU 18              *** External data addresses
I2       EQU 19              *
I3       EQU 20              *
I4       EQU 21              *
I5       EQU 22              *
I6       EQU 23              *
I7       EQU 24              *
I8       EQU 25              *
I9       EQU 26              ***
R1       EQU 27              *** Temporary locations
R2       EQU 28              *
R3       EQU 29              *
R4       EQU 30              *
R5       EQU 31              *
R6       EQU 32              *
R7       EQU 33              *
R8       EQU 34              *
R9       EQU 35              *
```

```
R10     EQU 36          *
S1      EQU 37          *
S2      EQU 38          *
S3      EQU 39          *
S4      EQU 40          *
S5      EQU 41          *
S6      EQU 42          *
S7      EQU 43          *
S8      EQU 44          *
S9      EQU 45          *
S10     EQU 46          *
T       EQU 47          *
T1      EQU 48          *
T2      EQU 49          *
T3      EQU 50          *
T4      EQU 51          *
U       EQU 52          *
U1      EQU 53          *
U2      EQU 54          *
U3      EQU 55          *
U4      EQU 56          ***
C72     EQU 57          *** Locations of coefficients
C73     EQU 58          *
C74     EQU 59          *
C71     EQU 60          *
C75     EQU 61          *
C76     EQU 62          *
C77     EQU 63          *
C78     EQU 64          *
C31     EQU 65          *
C32     EQU 66          *
C94     EQU 67          *
C92     EQU 68          *
C93     EQU 69          *
C98     EQU 70          *
C96     EQU 71          *
C97     EQU 72          ***
ONE     EQU 73          * Contains value 1
N1A     EQU 74          * N1 for the first module
N2A     EQU 75          * N2 for the first module
N1B     EQU 76          * N1 for the second module
N2B     EQU 77          * N2 for the second module
J       EQU 78          *
HOLDN   EQU 79          * Contains value N
UNSC    EQU 80          * Location of unscrambling constant
*
* Start of program memory section.
*
        AORG 0
START   LDPK 0
        LACK 1
        SACL ONE              * ONE = 1
        MPYK N
        PAC
        SACL HOLDN            * HOLDN = N
        MPYK CTABLE
        PAC                   * ACC has address of coefficient table
```

```
        LARK AR0,C72
        LARK AR1,15
RDCOEF    LARP 0                    * Read coefficients into data memory
          TBLR *+,AR1
          ADD  ONE
          BANZ RDCOEF
        LACK 16
        SACL UNSC                   * Unscrambling factor = 16 for N = 63
        LACK NI1                    * Initialize N1 and N2 for all modules
        SACL N1A
        SACL N2B
        LACK NI2
        SACL N2A
        SACL N1B
        ZAC
        SACL J                      * J = 0
*
* Module 1, WFTA = 7.
*
JLOOP1      LARK AR0,I1
            LARP 0                  * AR0 points to address locations
            LAC  J,1
            SACL *+                 * I1 = J (data organized as real value followed
*                                   * by imaginary so address I1 is 2 times J).
            LARK AR1,NI1-2
LLOOP1        LARP 0
              ADD  N2A,1
              SUB  HOLDN,1
              BGEZ MOD1
                ADD  HOLDN,1
MOD1          SACL *+,0,AR1         * I(L) = (I + N2)mod N
              BANZ LLOOP1           * Loop for L = 2 to N1
*
            CALL WFTA7              * Call WFTA module
            LAC  J
            ADD  N1A
            SACL J                  * J = J + N1
            SUB  HOLDN
            BLZ  JLOOP1             * Loop while J <= N
*
* Module 2, WFTA = 9.
*
        ZAC
        SACL J                      * J = 0
JLOOP2      LARK AR0,I1             * AR0 points to address locations
            LARP 0
            LAC  J,1
            SACL *+                 * I1 = J
            LARK AR1,NI2-2
LLOOP2        LARP 0
              ADD  N2B,1
              SUB  HOLDN,1
              BGEZ  MOD2
                ADD  HOLDN,1
MOD2          SACL *+,0,AR1         * I(L) = (I + N1)mod N
              BANZ LLOOP2           * Loop for L = 2 to N1
*
```

```
          CALL WFTA9                * Call WFTA module
          LAC  J
          ADD  N1B
          SACL J                    * J = J + N1
          SUB  HOLDN
          BLZ  JLOOP2               * Loop while J <= N
*
* Unscrambler.
*
        LARK AR0,62
        LARP 0
        ZAC
        SACL J
UNSCRA      OUT  J,PA0              * Read in data from address J
            IN   X1,PA1
            IN   Y1,PA1
            OUT  X1,PA2             * Output unscrambled data to port 2
            OUT  Y1,PA2
            LAC  J
            ADD  UNSC,1
            SUB  HOLDN,1
            BGEZ UNS1
               ADD HOLDN,1
UNS1        SACL J                  * J = (J + UNSC)mod N
            BANZ UNSCRA             * Loop over all data points
*
* PFA complete.
*
STOP B STOP
*********************************************************************
*
* WFTA modules.
*
*********************************************************************
*
* Length-7 module.
*
WFTA7   OUT  I1,PA0                 * Input 7 data points
        IN   X1,PA1
        IN   Y1,PA1
        OUT  I2,PA0
        IN   X2,PA1
        IN   Y2,PA1
        OUT  I3,PA0
        IN   X3,PA1
        IN   Y3,PA1
        OUT  I4,PA0
        IN   X4,PA1
        IN   Y4,PA1
        OUT  I5,PA0
        IN   X5,PA1
        IN   Y5,PA1
        OUT  I6,PA0
        IN   X6,PA1
        IN   Y6,PA1
        OUT  I7,PA0
        IN   X7,PA1
```

```
        IN   Y7,PA1
*
* Main length-7 calculations.
*
PFA7    LAC X2
        ADD X7
        SACL R1         * R1 = X2 + X7
        SUB X7,1
        SACL R2         * R2 = X2 - X7
        LAC Y2
        ADD Y7
        SACL S1         * S1 = Y2 + Y7
        SUB Y7,1
        SACL S2         * S2 = Y2 - Y7
        LAC X3
        ADD X6
        SACL R3         * R3 = X3 + X6
        SUB X6,1
        SACL R4         * R4 = X3 - X6
        LAC Y3
        ADD Y6
        SACL S3         * S3 = Y3 + Y6
        SUB Y6,1
        SACL S4         * S4 = Y3 - Y6
        LAC X4
        ADD X5
        SACL R5         * R5 = X4 + X5
        SUB X5,1
        SACL R6         * R6 = X4 - X5
        LAC Y4
        ADD Y5
        SACL S5         * S5 = Y4 + Y5
        SUB Y5,1
        SACL S6         * S6 = Y4 - Y5
        LAC R1
        ADD R3
        ADD R5
        SACL T1         * T1 = R1 + R3 + R5
        LAC S1
        ADD S3
        ADD S5
        SACL U1         * U1 = S1 + S3 + S5
        LAC X1
        ADD T1
        SACL X1         * X1 = X1 + T1
        LAC Y1
        ADD U1
        SACL Y1         * Y1 = Y1 + U1
*
        LT C71          * (+ -1.0)
        MPY T1
        PAC
        ADD X1,15
        SUB T1,15
        SACH T1,1       * T1 = X1 + C71 * T1
        MPY U1
        PAC
```

```
ADD Y1,15
SUB U1,15
SACH U1,1         * U1 = Y1 + C71 * U1
LT C72
MPY R1
PAC
MPY R5
SPAC
SACH T2,1         * T2 = C72 * (R1 - R5)
MPY S1
PAC
MPY S5
SPAC
SACH U2,1         * U2 = C72 * (S1 - S5)
LT C73
MPY R5
PAC
MPY R3
SPAC
SACH T3,1         * T3 = C73 * (R5 - R3)
MPY S5
PAC
MPY S3
SPAC
SACH U3,1         * U3 = C73 * (S5 - S3)
LT C74
MPY R3
PAC
MPY R1
SPAC
SACH T4,1         * T4 = C74 * (R3 - R1)
MPY S3
PAC
MPY S1
SPAC
SACH U4,1         * U4 = C74 * (S3 - S1)
LAC T1
ADD T2
ADD T3
SACL R1           * R1 = T1 + T2 + T3
LAC T1
SUB T2
SUB T4
SACL R3           * R3 = T1 - T2 - T4
LAC T1
SUB T3
ADD T4
SACL R5           * R5 = T1 - T3 + T4
LAC U1
ADD U2
ADD U3
SACL S1           * S1 = U1 + U2 + U3
LAC U1
SUB U2
SUB U4
SACL S3           * S3 = U1 - U2 - U4
LAC U1
```

```
        SUB U3
        ADD U4
        SACL S5          * S5 = U1 - U3 + U4
        LT C75
        MPY S2
        PAC
        MPY S4
        APAC
        MPY S6
        SPAC
        SACH U1,1        * U1 = C75 * (S2 + S4 - S6)
        MPY R2
        PAC
        MPY R4
        APAC
        MPY R6
        SPAC
        SACH T1,1        * T1 = C75 * (R2 + R4 - R6)
        LT C76
        MPY R2
        PAC
        MPY R6
        APAC
        SACH T2,1        * T2 = C76 * (R2 + R6)
        MPY S2
        PAC
        MPY S6
        APAC             * U2 = C76 * (S2 + S6)
        SACH U2,1
        LT C77
        MPY R4
        PAC
        MPY R6
        APAC
        SACH T3,1        * T3 = C77 * (R4 + R6)
        MPY S4
        PAC
        MPY S6
        APAC
        SACH U3,1        * U3 = C77 * (S4 + S6)
        LT C78
        MPY R4
        PAC
        MPY R2
        SPAC
        SACH T4,1        * T4 = C78 * (R4 - R2)
        MPY S4
        PAC
        MPY S2
        SPAC
        SACH U4,1        * U4 = C78 * (S4 - S2)
*
        LAC T1
        ADD T2
        ADD T3
        SACL R2          * R2 = T1 + T2 + T3
        LAC T1
```

```
        SUB T2
        SUB T4
        SACL R4         * R4 = T1 - T2 - T4
        LAC T1
        SUB T3
        ADD T4
        SACL R6         * R6 = T1 - T3 + T4
        LAC U1
        ADD U2
        ADD U3
        SACL S2         * S2 = U1 + U2 + U3
        LAC U1
        SUB U2
        SUB U4
        SACL S4         * S4 = U1 - U2 - U4
        LAC U1
        SUB U3
        ADD U4
        SACL S6         * S6 = U1 - U3 + U4
        LAC R1
        ADD S2
        SACL X2         * X2 = R1 + S2
        SUB S2,1
        SACL X7         * X7 = R1 - S2
        LAC S1
        ADD R2
        SACL Y7         * Y7 = S1 + R2
        SUB R2,1
        SACL Y2         * Y2 = S1 - R2
        LAC R3
        ADD S4
        SACL X3         * X3 = R3 + S4
        SUB S4,1
        SACL X6         * X6 = R3 - S4
        LAC S3
        ADD R4
        SACL Y6         * Y6 = S3 + R4
        SUB R4,1
        SACL Y3         * Y3 = S3 - R4
        LAC R5
        ADD S6
        SACL X5         * X5 = R5 + S6
        SUB S6,1
        SACL X4         * X4 = R5 - S6
        LAC S5
        ADD R6
        SACL Y4         * Y4 = S5 + R6
        SUB R6,1
        SACL Y5         * Y5 = S5 - R6
*
* Output resulting data values.
*
        OUT  I1,PA0
        OUT  X1,PA1
        OUT  Y1,PA1
        OUT  I2,PA0
        OUT  X2,PA1
```

```
        OUT  Y2,PA1
        OUT  I3,PA0
        OUT  X3,PA1
        OUT  Y3,PA1
        OUT  I4,PA0
        OUT  X4,PA1
        OUT  Y4,PA1
        OUT  I5,PA0
        OUT  X5,PA1
        OUT  Y5,PA1
        OUT  I6,PA0
        OUT  X6,PA1
        OUT  Y6,PA1
        OUT  I7,PA0
        OUT  X7,PA1
        OUT  Y7,PA1
*
        RET                        * End of WFTA7
*
* Length-9 module.
*
WFTA9   OUT  I1,PA0                * Input 9 data values
        IN   X1,PA1
        IN   Y1,PA1
        OUT  I2,PA0
        IN   X2,PA1
        IN   Y2,PA1
        OUT  I3,PA0
        IN   X3,PA1
        IN   Y3,PA1
        OUT  I4,PA0
        IN   X4,PA1
        IN   Y4,PA1
        OUT  I5,PA0
        IN   X5,PA1
        IN   Y5,PA1
        OUT  I6,PA0
        IN   X6,PA1
        IN   Y6,PA1
        OUT  I7,PA0
        IN   X7,PA1
        IN   Y7,PA1
        OUT  I8,PA0
        IN   X8,PA1
        IN   Y8,PA1
        OUT  I9,PA0
        IN   X9,PA1
        IN   Y9,PA1
*
* Main length-9 calculations.
*
PFA9    LAC X2
        ADD X9
        SACL R1             * R1 = X2 + X9
        SUB X9,1
        SACL R2             * R2 = X2 - X9
        LAC Y2
```

```
        ADD Y9
        SACL S1           * S1 = Y2 + Y9
        SUB Y9,1
        SACL S2           * S2 = Y2 - Y9
        LAC X3
        ADD X8
        SACL R3           * R3 = X3 + X8
        SUB X8,1
        SACL R4           * R4 = X3 - X8
        LAC Y3
        SUB Y8
        SACL S4           * S4 = Y3 - Y8
        ADD Y8,1
        SACL S3           * S3 = Y3 + Y8
        LAC X4
        ADD X7
        SACL R5           * R5 = X4 + X7
*
        LT C31
        MPY X7
        PAC
        MPY X4
        SPAC
        SACH T,1          * T = -(X4 - X7) * C31
        LAC Y4
        ADD Y7
        SACL S5           * S5 = Y4 + Y7
        MPY Y7
        PAC
        MPY Y4
        SPAC
        SACH U,1          * U = -(Y4 - Y7) * C31
*
        LAC X5
        ADD X6
        SACL R7           * R7 = X5 + X6
        SUB X6,1
        SACL R8           * R8 = X5 - X6
        LAC Y5
        ADD Y6
        SACL S7           * S7 = Y5 + Y6
        SUB Y6,1
        SACL S8           * S8 = Y5 - Y6
        LAC X1,15
        ADD R5,15
        SACH R9,1         * R9 = X1 + R5
        SUB R5,15
        LT C32
        MPY R5
        SPAC
        SACH T1,1         * T1 = X1 - R5 * C32
        LAC Y1,15
        ADD S5,15
        SACH S9,1         * S9 = Y1 + S5
        SUB S5,15
        MPY S5
        SPAC
```

```
        SACH U1,1        * U1 = Y1 - S5 * C32
*
        LT C92
        MPY R3
        PAC
        MPY R7
        SPAC
        SACH T2,1        * T2 = (R3 - R7) * C92
        MPY S3
        PAC
        MPY S7
        SPAC
        SACH U2,1        * U2 = (S3 - S7) * C92
        LT C93
        MPY R1
        PAC
        MPY R7
        SPAC
        SACH T3,1        * T3 = (R1 - R7) * C93
        MPY S1
        PAC
        MPY S7
        SPAC
        SACH U3,1        * U3 = (S1 - S7) * C93
        LT C94
        MPY R1
        PAC
        MPY R3
        SPAC
        SACH T4,1        * T4 = (R1 - R3) * C94
        MPY S1
        PAC
        MPY S3
        SPAC
        SACH U4,1        * U4 = (S1 - S3) * C94
*
        LAC R1
        ADD R3
        ADD R7
        SACL R10         * R10 = R1 + R3 + R7
        LAC S1
        ADD S3
        ADD S7
        SACL S10         * S10 = S1 + S3 + S7
        LAC T1
        ADD T2
        ADD T4
        SACL R1          * R1 = T1 + T2 + T4
        LAC T1
        SUB T2
        SUB T3
        SACL R3          * R3 = T1 - T2 - T3
        LAC T1
        ADD T3
        SUB T4
        SACL R7          * R7 = T1 + T3 - T4
        LAC U1
```

```
        ADD U2
        ADD U4
        SACL S1             * S1 = U1 + U2 + U4
        LAC U1
        SUB U2
        SUB U3
        SACL S3             * S3 = U1 - U2 - U3
        LAC U1
        ADD U3
        SUB U4
        SACL S7             * S7 = U1 + U3 - U4
        LAC R9,15
        ADD R10,15
        SACH X1,1           * X1 = R9 + R10
        SUB R10,15
        LT C32
        MPY R10
        SPAC
        SACH R5,1           * R5 = R9 - R10 * C32
        LAC S9,15
        ADD S10,15
        SACH Y1,1           * Y1 = S9 + S10
        SUB S10,15
        MPY S10
        SPAC
        SACH S5,1           * S5 = S9 - S10 * C32
*
        LT C31              * C31 = 136
        MPY R4
        PAC
        MPY R2
        SPAC
        MPY R8
        SPAC
        SACH R6,1           * R6 = -(R2 - R4 + R8) * C31
        MPY S4
        PAC
        MPY S2
        SPAC
        MPY S8
        SPAC
        SACH S6,1           * S6 = -(S2 - S4 + S8) * C31
        LT C96              * C96 = 142
        MPY R4
        PAC
        MPY R8
        APAC
        SACH T2,1           * T2 = (R4 + R8) * C96
        MPY S4
        PAC
        MPY S8
        APAC
        SACH U2,1           * U2 = (S4 + S8) * C96
        LT C97
        MPY R2
        PAC
        MPY R8
```

```
        SPAC
        SACH T3,1          * T3 = (R2 - R8) * C97
        MPY S2
        PAC
        MPY S8
        SPAC
        SACH U3,1          * U3 = (S2 - S8) * C97
        LT C98             * C98 = 141
        MPY R2
        PAC
        MPY R4
        APAC
        SACH T4,1          * T4 = (R2 + R4) * C98
        MPY S2
        PAC
        MPY S4
        APAC
        SACH U4,1          * U4 = (S2 + S4) * C98
*
        LAC T
        ADD T2
        ADD T4
        SACL R2            * R2 = T + T2 + T4
        LAC T
        SUB T2
        SUB T3
        SACL R4            * R4 = T - T2 - T3
        LAC T
        ADD T3
        SUB T4
        SACL R8            * R8 = T + T3 - T4
        LAC U
        ADD U2
        ADD U4
        SACL S2            * S2 = U  + U2 + U4
        LAC U
        SUB U2
        SUB U3
        SACL S4            * S4 = U -  U2 - U3
        LAC U
        ADD U3
        SUB U4
        SACL S8            * S8 = U  + U3 - U4
        LAC R1
        ADD S2
        SACL X9            * X9 = R1 + S2
        SUB S2,1
        SACL X2            * X2 = R1 - S2
        LAC S1
        ADD R2
        SACL Y2            * Y2 = S1 + R2
        SUB R2,1
        SACL Y9            * Y9 = S1 - R2
        LAC R3
        ADD S4
        SACL X3            * X3 = R3 + S4
        SUB S4,1
```

```
        SACL X8           * X8 = R3 - S4
        LAC S3
        ADD R4
        SACL Y8           * Y8 = S3 + R4
        SUB R4,1
        SACL Y3           * Y3 = S3 - R4
        LAC R5
        ADD S6
        SACL X7           * X7 = R5 + S6
        SUB S6,1
        SACL X4           * X4 = R5 - S6
        LAC S5
        ADD R6
        SACL Y4           * Y4 = S5 + R6
        SUB R6,1
        SACL Y7           * Y7 = S5 - R6
        LAC R7
        ADD S8
        SACL X6           * X6 = R7 + S8
        SUB S8,1
        SACL X5           * X5 = R7 - S8
        LAC S7
        ADD R8
        SACL Y5           * Y5 = S7 + R8
        SUB R8,1
        SACL Y6           * Y6 = S7 - R8
*
* Output resulting data values.
*
        OUT  I1,PA0
        OUT  X1,PA1
        OUT  Y1,PA1
        OUT  I2,PA0
        OUT  X2,PA1
        OUT  Y2,PA1
        OUT  I3,PA0
        OUT  X3,PA1
        OUT  Y3,PA1
        OUT  I4,PA0
        OUT  X4,PA1
        OUT  Y4,PA1
        OUT  I5,PA0
        OUT  X5,PA1
        OUT  Y5,PA1
        OUT  I6,PA0
        OUT  X6,PA1
        OUT  Y6,PA1
        OUT  I7,PA0
        OUT  X7,PA1
        OUT  Y7,PA1
        OUT  I8,PA0
        OUT  X8,PA1
        OUT  Y8,PA1
        OUT  I9,PA0
        OUT  X9,PA1
        OUT  Y9,PA1
*
```

```
        RET             * End of length-9 module
*
*********************************************************
*
* Coefficient table.
*
*********************************************************
CTABLE EQU $
CFC72   DATA >6523      * C72 = 0.79015647
CFC73   DATA >0726      * C73 = 0.055854267
CFC74   DATA >5DFD      * C74 = 0.7343022
CFC71   DATA >EAAA      * C71 = -0.1666667 (+ -1.0 IN THE PROGRAM)
CFC75   DATA >3871      * C75 = 0.44095855
CFC76   DATA >2BA1      * C76 = 0.34087293
CFC77   DATA >4459      * C77 = 0.53396936
CFC78   DATA >6FFA      * C78 = 0.87484229
*
CFC31   DATA >6ED9      * C31 = 0.86602540
CFC32   DATA >4000      * C32 = 0.50000000
*
CFC94   DATA >620D      * C94 = 0.76604444
CFC92   DATA >7847      * C92 = 0.93969262
CFC93   DATA >E9C5      * C93 = -0.17364818
CFC98   DATA >ADB9      * C98 = -0.64278761
CFC96   DATA >D438      * C96 = -0.34202014
CFC97   DATA >81F1      * C97 = -0.98480775
*
        END
```

```
        IDT 'CONV'
************************************************************************
*
*       A general routine for a length-N linear convolution.
*       For this particular implementation, N = 32 and implements
*       a bandpass FIR filter with linear phase.
*
************************************************************************
*
XNEW    EQU 0           * Newest input sample (always at location 0)
X       EQU 31          * End of data points X
H       EQU 63          * End of impulse response sequence
YOUT    EQU 64          * Output location
ONE     EQU 65          * Contains the value 1
*
        AORG 0
        B START         * Branch to the beginning of the program
*
* Impulse response terms.
*
H1      DATA >02C0
H2      DATA >00E9
H3      DATA >FFC6
H4      DATA >01B5
H5      DATA >FFCD
H6      DATA >FA22
H7      DATA >FBC3
H8      DATA >0380
H9      DATA >03A5
H10     DATA >FFE6
H11     DATA >0694
H12     DATA >0AB0
H13     DATA >F6A8
H14     DATA >E250
H15     DATA >F6BA
H16     DATA >1F1C
H17     DATA >1F1C
H18     DATA >F6BA
H19     DATA >E250
H20     DATA >F6A8
H21     DATA >0AB0
H22     DATA >0694
H23     DATA >FFE6
H24     DATA >03A5
H25     DATA >0380
H26     DATA >FBC3
H27     DATA >FA22
H28     DATA >FFCD
H29     DATA >01B5
H30     DATA >FFC6
H31     DATA >00E9
H32     DATA >02C0
*
* Begin Program.
*
START   LDPK 0
        LACK 1
```

```
        SACL ONE        * ONE = 1
*
        LARK AR0,H      * AR0 addresses data locations
        LARK AR1,31     * AR1 is used as a loop counter
        LACK H32
LOADH   LARP AR0        * Load the impulse response
        TBLR *-,AR1
        SUB ONE
        BANZ LOADH
*
        LARK AR1,X      * AR1 used to address data and as a counter
        ZAC
LOADX   SACL *          * Initialize filter
        BANZ LOADX
*
        LARP AR0
NXTPT   IN XNEW,PA0     * Get next input sample
*
        LARK AR0,X      * AR0 points to the input sequence
        LARK AR1,H      * AR1 points to the impulse response
*
        ZAC
        LT *-,AR1
        MPY *-,AR0
LOOP    LTD *,AR1       * Load and move input sequence, accumulate result
        MPY *-,AR0      * Multiply impulse response
        BANZ LOOP       * Loop N times
*
        APAC            * Accumulate last multiply
        SACH YOUT,1
        OUT YOUT,PA1    * Output accumulated result
*
        B NXTPT         * Get the next input sample
        END
```

Chapter 6

COMPARISONS AND CONCLUSIONS

6.1 INTRODUCTION

The comparison of various digital signal processing algorithms is complicated because it not only depends on the application, but also on the hardware/software used and on the ability and experience of the programmer. In this chapter are presented some general guidelines and some specific comparisons that should be used as a starting point in choosing an appropriate algorithm.

In order to get some feeling for the relative merits of the various FFT algorithms, the execution times for several of the FORTRAN programs from Chapter 4 are presented. These depend on the compiler and computer, but do give some general measure of the usefulness of a particular approach. The timing of several of the TMS32010 assembly language programs is also given and discussed.

6.2 TIMINGS OF THE FORTRAN PROGRAMS

Most of the timings are for a PDP 11/45 with hardware floating-point arithmetic having about equal addition and multiplication times. Table 6-1 gives the execution times in milliseconds for complex data of lengths 64, 256, and 1024. For the PFA and WFTA, the lengths are 63, 252, and 1008.

TABLE 6-1. TIMINGS OF FORTRAN FFT PROGRAMS
(time in milliseconds on a PDP 11)

EXAMPLE PROGRAM	CHAPTER 4 PROGRAM NO.	LENGTH-N 64/63	256/252	1024/1008
Radix-2 FFT				
Calculate TFs, 1BF	5	93	434	1968
Table of TFs, 1BF	6	65	316	1480
Table of TFs, 2BFs	9	61	301	1435
Table of TFs, 3BFs	10	60	295	1400
Radix-4 FFT				
Calculate TFs, 1BF	11	88	418	1908
Table of TFs, 1BF		59	293	1375
Table of TFs, 2BFs		55	277	1332
Table of TFs, 3BFs	12	54	273	1320
Radix-8 FFT				
Calculate TFs, 1BF	15	82		
Table of TFs, 2BFs	16	48		
PFA1 with unscrambler	17	54	254	1190
PFA2 in-place, in-order	19	48	228	1086
WFTA		72	306	1590

The purpose of this table is to show the effects of using radix 2, 4, and 8 for the FFT; of using table lookup rather than calculation of the cosines and sines; of using one, two, or three butterflies in the FFT; and for using PFA and WFTA.

A large improvement of 20 to 30 percent is obtained by using a table lookup of the twiddle factors; however, the table requires 2N memory locations. There is approximately a 10 percent improvement by using radix-4 rather than radix-2, and a 7 percent improvement by using radix-8 rather than radix-4. The use of two butterflies rather than one gives a greater improvement than using three rather than two. Although there are fewer floating-point operations in a five-butterfly program, a FORTRAN program of the type presented in Chapter 4 using five butterflies runs slower and thus less efficiently than a three-butterfly program. The PFA2 is faster than the various FFTs while the WFTA is slower than the table lookup FFTs and faster than the FFTs that calculate the TFs.

Program 14 is approximately 8 percent faster than Program 12, and Program 13 is somewhere between Programs 12 and 14 in speed. The improvements in speed for Programs 13 and 14 are even more dependent on the computer and compiler than for the others. Program 8, which uses an up-date method to calculate the twiddle factors, is only 6 to 8 percent slower than Program 6, but as was discussed in Chapter 4, it has more quantization error than either Program 5 or 6.

These times should be compared to the operation counts given in Tables 2-4, 2-6, and 2-8. The number of data multiplications and additions (floating-point in the FORTRAN case) tells part of the story, but not all. Time is also required for calculating the indices, transferring the data, calculating the TFs, and unscrambling the DFT order. It is informative to investigate the relative time required for these operations in a particular algorithm.

Table 6-2 gives the percentage of the total execution time required by each of the internal sections of a typical program in three classes of algorithms. The first class contains the Cooley-Tukey FFTs that calculate the TF values, such as in Program 5 or 7 in Chapter 4. The second class contains the FFTs that use a table for the TFs such as in Program 6, and the third class contains the PFA of Program 8. This table gives an indication of where effort should be expended in trying to improve efficiency. It shows why using table lookup of the TFs improves the FFT. It also indicates the possible speed increases that could be obtained removing the unscrambler. The PFA is particularly efficient in that it spends most of its time on data arithmetic and little on overhead. Although not analyzed in this table, the WFTA uses a large percentage of its execution time on indexing. This explains its relatively slow execution times in Table 6-1, even though the number of its data operations is comparable to those for the FFT and PFA.

TABLE 6-2. PERCENTAGE OF EXECUTION TIME FOR THE FORTRAN PROGRAMS

SECTION OF PROGRAM	TYPE OF PROGRAM		
	FFTs WITH CALCULATED TFs	FFTs WITH TF TABLES	PFA
Indexing	13	17	9
Unscrambling	7	8	8
Calculate TFs	20		
TF multiply	30	37	
Butterfly	30	38	83

6.3 TIMINGS OF THE TMS32010 ASSEMBLY LANGUAGE PROGRAMS

Because the instruction set of the TMS32010 has some characteristics that are somewhat different from those of a general-purpose computer, the results of FORTRAN program evaluation cannot always be used without further examination. Table 6-3 gives timings for the TMS32010 assembly language programs. In addition to the FFTs and PFA analyzed in Table 6-1, a direct DFT program and a Goertzel algorithm program are timed. The Goertzel algorithm is approximately twice as fast as the direct calculation. This is in agreement with the operation count in Section 2.2.3 which showed that the Goertzel algorithm required half the number of multiplications of the direct calculation.

The improvement of the FFT over the direct and Goertzel algorithms is also predicted fairly well by the operation counts. The relative times of the one-, two-, and three-butterfly FFTs, the radix-2, -4, and -8, and the PFA are in approximate agreement with the operation counts in Chapter 2 and with the FORTRAN program times in Table 6-1. However, the assembly language unscramblers take a significantly larger percentage of the total execution time than the FORTRAN versions. The assembly language unscrambler for the FFT with N = 64 is about 0.47 ms, and with N = 1024 about 7.4 ms. For PFA1, the times are 0.23 ms for N = 63 and 1.8 ms for N = 504.

TABLE 6-3. TIMINGS OF ASSEMBLY LANGUAGE PROGRAMS (time in milliseconds on the TMS32010)

EXAMPLE PROGRAM	CHAPTER 5 PROGRAM NO.	LENGTH-N 64/63	512/504	1024/1008
Direct DFT	1	29		
Goertzel	2	14		
FFT radix-2 1BF	3	2.87		69.4
FFT radix-4 1BF	4	1.92		45.3
FFT radix-4 3BF	5	1.72		42.3
FFT radix-4 3BF straight-line	6	0.60		
FFT radix-8 2BF	7	1.41	16.0	
PFA1	8	1.53	17.4	
PFA2	8	1.30	15.6	
PFA2 straight-line		0.83		

A very interesting result, illustrated in Table 6-3, is the efficiency of straight-line code on the TMS32010. The execution of the straight-line code of the radix-4 FFT in Program 6 of Chapter 5 is more than twice as fast as the looped version in Program 5. This is explained by the fact that a test and branch instruction takes as much time as a multiply and accumulate on the TMS32010. The improvement for the PFA is not as great, but still significant. Indeed, the use of straight-line code rather than loops is almost always much more important on the TMS32010 than for traditional computer or microprocessor implementations. A compromise of partial straight-line code, as illustrated in Figure 2-4, can give almost as much speed with much shorter total program length.

Although not timed in Table 6-3, Rader's method of Section 2.2.5 would probably be faster than the Goertzel algorithm and with less quantization error.

6.4 CONCLUSIONS

From the results of this chapter, a general approach to the development of a digital signal processing algorithm can be formulated. First, a variety of possibilities should be considered based on the fundamental theory presented in Chapters 2 and 3. Next, a high-level language implementation of the most promising candidates would be made and tested. Finally, an assembly language program would be written and tested, keeping in mind the special characteristics of the TMS32010 instruction set. This development procedure would probably be repeated several times taking into account other factors such as memory constraints, input-output requirements, and quantization effects.

From the analysis in Chapter 2 and the timings in this chapter, it was concluded that the radix-2, radix-4, and PFA programs gave fast execution, reasonable length code, and a variety of data lengths. The mixed-radix FFT and WFTA did not seem as well suited to most general-purpose computers or to the TMS32010. If speed is very important, then the use of radix-8, multiple butterflies, straight-line code, and multiply-immediate instruction should all be considered, but each has its own disadvantage.

If lengths not available with the basic algorithms from Chapter 2 are desired, mixed versions can be constructed using the index map methods from Section 2.2.7. For example, by following a radix-8 FFT with one stage of radix-2 or radix-4, any length that is a power of two can be processed. A radix-2 FFT could be used as one module in a PFA to give a very large set of possible lengths.

If only a few DFT values are needed, use of the direct method, Goertzel's algorithm, or Rader's method is probably faster than any FFT or PFA program.

For digital convolution, the fastest approach is a direct calculation until the filter length exceeds around 30 for a FORTRAN program run on a general-purpose computer. For longer length convolution, the use of data segmentation and the FFT or PFA becomes faster, but at the expense of a more complicated program. The exact filter length where transforms become more efficient depends on which FFT or PFA is used, on the block lengths, on whether the data are real or complex, and on other factors that are specific to an application. For the TMS32010, the theoretical crossover length is around 100, but because of the limited memory, the use of an FFT approach may not be practical.

It has been the goal of this book to open the possibility of using a variety of modern, practical digital signal processing algorithms. The reader still has the task of experimentation and evaluation for a particular application.

INDEX

T

U

W

Z